A Matter of Wonder

GOTTFRIED SCHATZ

A MATTER
OF WONDER

WHAT BIOLOGY REVEALS
ABOUT US, OUR WORLD,
AND OUR DREAMS

Translated by Andrew Shields

 Basel · Freiburg · Paris · London · New York · New Delhi · Bangkok ·
Beijing · Tokyo · Kuala Lumpur · Singapore · Sydney

TABLE OF CONTENTS

FOREWORD

It is a great pleasure for me to provide a few words of introduction to the English edition of the essays of Gottfried 'Jeff' Schatz, an outstanding biologist and biochemist, a man who has dedicated his life to science but doesn't mind standing at a critical angle to it. His essays deal with key issues in the natural sciences while extending his thought beyond them to universal concerns. They are models of penetrating free thought.

As these essays make plain, science is no soulless enterprise, but a way of humanizing the universe, of taking steps toward making it more familiar and understandable. Passion, courage and patience: these are the characteristics of great scientists. A scientific way of thinking embodies basic trust that the laws of nature are understandable, combined with open-minded skepticism with regard to any particular conclusion about them. It proceeds by finding significance in what had previously gone unnoticed, by forming and testing hypotheses, and by providing evidence to support its claims. It is a powerful, tentative, unending enterprise.

Jeff, a person steeped in humanistic culture, emphasizes the commonalities between scientific and artistic endeavor. He is a trenchant defender and critic of our scientific establishment and our universities, always insisting that the purpose of those institutions is to foster active intellectual inquiry, creative thinking and new discoveries. An eloquent advocate of basic research, he argues that the connection between science and technology can be enlivening and enriching for both, as is true for the connection between art and craft.

To Heimo – again

Translated and adapted from *Jenseits der Gene – Essays über unser Wesen, unsere Welt und unsere Träume*, NZZ Libro, Zürich, 2008

Library of Congress Cataloging-in-Publication Data

Schatz, G. (Gottfried)
    [Jenseits der Gene. English]
    A matter of wonder : what biology reveals about us, our world, and our dreams / Gottfried Schatz ; translated by Andrew Shields.
        p. cm.
    ISBN 978–3–8055–9744–9 (hard cover : alk. paper)
    1. Biology––Philosophy. I. Title.
    QH331.S2813 2011
    570.1––dc22

                                                                2011009950

© Copyright 2011 by S. Karger AG,
P.O. Box, CH–4009 Basel (Switzerland)
Printed in Switzerland on acid-free and non-aging paper
(ISO 9706) by Reinhardt Druck, Basel
ISBN 978–3–8055–9744–9

*Few people have the ability to communi-
cate the essence and the beauty of science.
One, already known to the international public, is the French Nobelist François Jacob. Another is Jeff Schatz, the author of this memo-
rable, wise book.*

Fotis Kafatos, PhD, FoMeRS

*Professor, Immunogenomics Chair,
    Imperial College London
Adjunct Professor,
    Harvard School of Public Health
Honorary and Founding President,
    European Research Council*

STRANGERS WITHIN ME

STRANGERS WITHIN ME

Never had I been so cold. On the run from the chaos of war, we found ourselves stuck on an unheated train in the middle of nowhere on a February night so bitter cold that sleep was impossible. Only at daybreak did I secretly snuggle up against the man dozing beside me; the warmth of his body finally gave me the sleep I had been longing for. I will never forget that comforting warmth – but where did it come from? I could not have suspected that the heat of that man's body would one day become the focus of my research and would tell me a story from the dawn of life itself.

The cells of my body burn food to generate energy. In this 'cell respiration', they consume oxygen gas and capture part of the energy released by the combustion as useful chemical energy. The harder a cell has to work, the faster it respires. Brain cells respire faster than all other body cells and, relative to their weight, produce ten thousand times more energy per second than the sun.

I owe all this to tiny combustion machines inside my cells – the mitochondria. Under the microscope, they usually look like little worms, but they can also form a connected net running through a whole cell. They have their own DNA, which carries the blueprint for thirteen proteins, each of them an indispensable part of the combustion machine. If just one of these proteins is defective or, worse, completely missing, the result can be blindness, deafness, muscular atrophy, dementia, or early death. Why do my mitochondria have a few genes of their own, while all my other genes are stored in the chromosomes in the cells' nucleus? The explanation is a matter not of logic but of history, going back to the origins of life – and no story is as magnificent and exciting.

There have been living cells on our planet for at least 3.8 billion years. The first organisms probably drew their energy from fermentation – just like today's yeast cells, which split sugar into alcohol and carbon dioxide. Fermentation does not release much energy, but neither does it need any oxygen gas. This was crucial for early cells, as the earth's atmosphere contained no oxygen back then. As such organisms multiplied and used up the fermentable matter, they drifted into a dangerous energy crisis. Salvation came from a new kind of organism that fed on sunlight, thus tapping into an almost unlimited energy source for life on Earth – the fusion of atomic nuclei in the sun. These light-harvesting cells soon ran rampant over the globe; even today, gigantic mounds of petrified fossils bear witness to them.

However, the use of sunlight released oxygen gas from water, and oxygen causes oxidation that is harmful to cells. The resulting oxygen poisoning may well have precipitated the greatest mass extinction in the history of life. Poisonous waste, though, always stimulates the inventiveness of evolution – and so it did not take long for cells to find a way to use oxygen gas to burn the organic remains of other cells and to live on the energy of that combustion. Cell respiration entered the scene. Although these respiring cells lived as parasites on the remains and the waste of other cells, they were very successful, as they could also grow at night and in places sunlight does not reach.

About two billion years ago, then, our planet harbored three main types of organisms, all of them similar to today's bacteria. All three had only a small amount of DNA, so they did not have enough biological information to form complex, multicellular organisms. The first fed on the energy of sunlight. The second burned the remains of dead organisms. And the third did neither; like the earliest organisms, it lived as best as it could on the fermentation of organic matter. And yet it was this third kind, this throwback, that made a dazzling breakthrough: it trapped respiring organisms and used them as energy producers, in exchange for a protective environment and more efficient safekeeping of their DNA. This symbiosis apparently made both partners happy and has survived until today. It created a new kind of cell that could not only generate plenty of energy, but also combined the genetic information of two organisms. Now Nature finally had the material to build complex plants and animals. In the course of the next 1.5 billion years, the entrapped respiring bacteria adapted to their protective environment. They surrendered more and more of their DNA to their host and soon could no longer live on their own. They became their host's respiratory organs – mitochondria. I am the distant descendant of a deal made 1.5 billion years ago between two different organisms – and the DNA inside my mitochondria is the meager remnant of the DNA of those once free bacteria.

As they evolved for a sedentary life inside their host cells, the respiring bacteria took over so many important metabolic processes in their host that the latter could no longer live alone either. When the cells in my body divide, they bequeath to their daughter cells not only the DNA in the chromosomes of the cell's nucleus, but also the mitochondria – along with *their*

DNA. Fathers play no role in the passing on of mitochondria, for the mitochondria of a fertilized egg cell come only from the mother. Sperm cells have only one single mitochondrion that probably cannot penetrate the egg cell – and if it did manage to do so, it would have no chance against the hundreds of thousands of maternal mitochondria in the much larger egg cell.

Mitochondria tend their fire with great care and damp it when the cell has enough energy. If this regulation fails, catastrophe strikes. A tragic example of such failure was observed in 1959 when a 27-year-old Swedish woman sought help in a clinic because she kept sweating even on the coldest days and remained extraordinarily thin despite her voracious appetite. The doctors recognized that her mitochondrial fires were burning out of control, but could not help her. Ten years later she took her own life in despair.

Even in healthy mitochondria, combustion is not perfect and generates highly oxidizing waste products that can destroy vital cell molecules. One of these is DNA, in which almost any chemical change can cause harmful or even deadly mutations. Chemical 'sparks' from my mitochondrial fires damage my DNA hundreds – or even thousands – of times every day, but fortunately, my cells can usually quickly undo the damage. Still, a few mutations do remain, above all those in the DNA of my mitochondria, and they accumulate over the years, gradually but inexorably. As a result, my mitochondria have more and more trouble forming the thirteen proteins of their combustion machine from their own DNA, so they give my cells less and less energy. The harmful side effects of my already wheezing cell respiration are in part responsible for the aging not only of my mitochondria but of my whole body. Should I be lucky – or unlucky – enough to live to be ninety, every single one of my muscle cells will have to get by with up to ten times less energy than in my youth. But it is my brain and its outward extensions – the retinas of my eyes – that face the worst prospects. Their mitochondrial fires burn especially fast, so they feel the full force of any imperfections in my mitochondria and age more quickly than other tissues in my body. Parkinson's disease, Alzheimer's disease, degeneration of the retina – all these terrifying diseases of old age are caused in part by chemical sparks from worn-out mitochondria.

As mitochondria age, they emit more and more harmful waste that in turn accelerates the aging process even more. For a cell, the only way out of this vicious circle is often suicide. When mitochondria are so badly damaged that they can no longer meet the cell's energy needs, they emit chemical signals that order the cell to kill itself. The cell then digests itself, packs its remains into little membrane sacks and leaves these as prey for roving scavenger cells. This cellular hara-kiri produces neither messy waste nor inflammation of the surrounding tissue – the cell orchestrates its suicide as carefully as it did its growth and division. What could provide more vivid proof that life and death are inseparable parts of a greater whole? Just as Persephone, the daughter of the fertility goddess, rules beside Hades over the dead, the mitochondria that give my cells life can also herald their death. The respiring bacteria and their hosts have not yet fully come to terms with each other. They are still trying to work things out. Mitochondria may be part of me, but they are still strangers.

OMINOUS VISITORS

OMINOUS VISITORS

We humans have not one but two hereditary systems – one chemical and one cultural. The chemical system is based on strands of DNA molecules and other components of our cells; it determines what we can be. The cultural system consists of the dialogue between generations; it determines what we then become. Our chemical system barely sets us apart from animals, but our cultural system has no equal in nature. Its productive power provides us with language, art, science, and ethics. In both of these systems, the transmission of information from one generation to the next is extremely reliable, but each does make occasional mistakes. In the chemical system, mistakes in transmission (mutations) change our bodies, while mistakes in the cultural system change our behavior. In the long run, such mistakes protect us from biological and cultural stagnation, but in the short run, they can be catastrophic. In the early Middle Ages, the Tay-Sachs mutation in the chemical system of an Eastern European Ashkenazim caused his brain to wither, and many of his contemporary descendants now face the same fate. And the twentieth century reminded us again of the horrors of cultural mutations.

Which of these two hereditary systems leads people from different cultures to think and act so differently? Perhaps it is sometimes neither, but instead a parasite that takes up residence in our brains.

There is no question that parasites can change the behavior of animals. When certain nematodes infect land-living locusts or crickets, they secrete proteins and other neuroactive substances that change the insects' sense of balance, and probably other brain functions as well. As soon as a nematode is fully grown and sexually mature inside its insect, the latter loses its fear of water, throws itself suicidally into the nearest pond and, while fighting for its life, releases its worm, which by this point is almost three times as long as its host. The parasite immediately swims away to look for a mate. And when the larvae of a trematode infect a Pacific killifish, the fish abandons its innate caution, capering wildly and twisting around on the surface of the ocean until it attracts the attention of birds of prey. As a result, the birds eat an average of about thirty times more infected fish than healthy fish. This brainwashing is a necessary part of the trematode's life cycle, which needs three different hosts. In the birds' intestines, the worm produces its eggs, which are then excreted in salt marshes on the California coast and eaten by snails, in which they develop into larvae. The larvae then infect more killifish and finally return with them into bird intestines.

The cases of intelligent mammals, such as mice and rats, are even more striking. When the single-celled *Toxoplasma gondii* infects them, it tries to settle in those regions of the brain that govern emotions and fear, which turns the rodents' innate fear of cat odor into its opposite – a fatal attraction. This makes it much more likely, of course, that the infected animals will fall prey to cats – which is precisely where the parasite needs to go, for only in the intestines of cat species can *Toxoplasma gondii* form its egg-like oocysts, which then have to find their way into a warm-blooded intermediate host, such as a rat. The brainwashing of the rodents is very precise, for the parasite does not affect their innate fear of open spaces or unknown food.

Even humans can be intermediate hosts for *Toxoplasma gondii*. Indeed, billions of us are, because we eat unwashed vegetables or raw meat infected with oocysts, or because we are not aware that our cute house cat can give us the oocysts. In Great Britain a few years ago, almost forty percent of all meat products available for sale contained *Toxoplasma gondii* genes, and the percentage is surely even higher in many poorer countries. So it is not surprising that about a third of all North Americans and almost half the population of Switzerland carry antibodies against the parasite in their blood – incontrovertible evidence that they were once infected, or even still are. Many infections are not identified as such; we can remain infected for the rest of our lives without any noticeable symptoms. However, in pregnant women who are not already immune to the parasite, an infection can deform or even kill the embryo. Further, such infections might even trigger schizophrenia in some people – in fact, some medications for schizophrenia are also effective in curing *Toxoplasma gondii* infections. These days, we hardly ever fall prey to big cats anymore, so the parasite no longer gains any noticeable advantage from infecting humans. But *Toxoplasma* might still alter our minds in subtle ways: preliminary and still unconfirmed studies suggest that infected women tend to be more intelligent, dynamic, and independent, while infected men are more jealous, conservative, and group conscious. In both sexes, the parasite makes us more likely to feel guilty, which many psychologists consider a negative emotional disposition.

Have parasites played a role in shaping the character of human cultures? If *Toxoplasma gondii* actually does make men more conscious of tradition and more loyal to their group, it could also be partly responsible for the way in which some cultures more stubbornly defend traditional gender roles than others do or that some of them value ambition and material success more than intellectual depth and human relationships. And could it be that reduced openness to new things and new ideas has reduced the innovative power of entire cultures? Extensive surveys in thirty-nine countries do in fact suggest that the negative emotional disposition of a population correlates with the extent of *Toxoplasma gondii* infection. Of course, the idea that cultural idiosyncrasies are not the result but the cause of the infection cannot be completely dismissed. However, the weight of the evidence makes this unlikely, so studies of the role of parasites in the development of human cultures could still provide a few surprises.

The idea that parasites might play a role in what I think and do flies in the face of my sense of myself and my image of humans. Can I still sing the song *Die Gedanken sind frei* ('thoughts are free') as innocently as I did when I was a child? Or should I try to overcome my scientific perspective and feel the wholeness of nature, as artists and mystics often do? From this perspective, parasites that change how I think might only be an especially marvelous example of the unity of life's web on our blue planet. Human reason does give us the weapons to detect and destroy such parasites. Yet who will protect us from the immaterial parasites that take over our thoughts and emotions? They are numerous – racism, religious fanaticism, national hysteria, spiritualism, and superstition. They are highly contagious, and they dehumanize us far more than *Toxoplasma gondii* ever could. As long as we have not learned to recognize these ominous visitors in time and do not combat them effectively, they will be our greatest danger.

PUNISHING PROTECTORS

PUNISHING PROTECTORS

Throughout our lives, we long for the feeling of security we had as children. Could this have inspired Thomas Wolfe to entitle his great epic *Look Homeward, Angel*? But what has become of the guardian angel that used to watch over me? My parents and teachers, who gave me that angel, took him with them into the grave. Even today, though, numerous protectors still keep me safe. They are my body's sensors, which let me see, hear, smell, taste – and feel pain.

Nothing alerts me to danger as effectively as pain. It keeps me from scalding myself on hot tea, stepping barefoot on a sharp stone, or moving a broken leg. And it raises its warning voice when something inside my body is amiss. The pain sensors on the surface of my body are thickly sown; they react in split seconds to locate the source of a pain to the nearest millimeter. The sensors inside my body work more slowly; they are less densely distributed and often do not tell me exactly where a pain comes from. They can even mislead me, disguising a heart problem as harmless shoulder pain. As for my brain, it cannot feel any pain at all.

Whoever feels no pain lives dangerously – and often very briefly. In the first half of the twentieth century, doctors identified patients who were completely insensitive to pain. A few years ago, a research team working in northern Pakistan even discovered a whole family whose members did not know what pain meant. As children, many of them had bitten off parts of their tongues or broken limbs without realizing what had happened. One boy earned money as a performer by walking on glowing coals or stabbing his own arm with a knife. He died just before his fourteenth birthday when he jumped off a roof. Such people may perceive the cut of a knife, but not the attendant pain that the rest of us would feel.

They are completely healthy – all that they are missing is one intact protein that helps pain-sensitive nerve cells emit electrical signals. Many people have an overactive variant of this protein; their brains are overrun by erroneous signals that make them feel intense, often unbearable pains. Their protective guardian angel has turned into a cruel torturer.

In our fight against pain, we may have won a few battles, but we have not yet won the war. In our complex bodies, unconsciousness and death are dangerously close neighbors, and consequently many a pain-relieving potion has proven lethal. The brilliant Paracelsus recognized the anesthetic power of ether in the middle of the sixteenth century, but he did not come up with the idea of using it to relieve pain during operations. People thus still had to suffer the horrors of unanesthetized surgery for almost three centuries until, on October 13, 1804, the Japanese doctor Seishu Hanaoka used general anesthesia while removing a breast tumor from a 60-year-old woman. At the time, though, the Tokugawa shogunate had sealed off Japan from the outside world, and so this magnificent achievement remained as unknown in the West as the composition of the plant extract Hanaoka had used as the anesthetic. It was not until 1841 that the 27-year-old provincial American doctor Crawford Williamson Long began to anesthetize his patients with ether before operations. But since he only published his successful results six years later, the ambitious and industrious William T. G. Morton was long considered the inventor of ether anesthesia. Ether was soon followed by chloroform and eventually by a rich palette of ever more effective and ever safer anesthetic gases; only rarely do they still cause death. We still do not know exactly how these gases work; they probably bind to water-repellent niches in the protein molecules that act as pain sensors, thus suppressing their function.

Our fight against pain, though, is more than just a battle against the complexity of the human body; it is also a battle against human folly. For many, pain is God-given, and its suppression a sin. After all, the Book of Genesis is cruelly clear: 'Unto the woman he said, I will greatly multiply thy sorrow and thy conception; in sorrow thou shalt bring forth children.' The literal interpretation of this fateful passage led to fierce opposition very soon after the first successful uses of ether anesthesia. In 1865, the city fathers of Zurich forbade this method with the argument that pain is a natural, divinely intended punishment for original sin, making every attempt to eliminate it unethical. In 1853, when Queen Victoria was given chloroform during the birth of her eighth child, even the prestigious British scientific journal *The Lancet* was shocked. Such reservations now belong to the past. For me, pain-free surgery is the most inspiring and humane technical invention of the past two millennia. A few years ago, when I was lying on an operating table with appendicitis and the anesthesiologist bent down over me, I suddenly saw the guardian angel of my childhood. Secretly, I hoped that he would keep my self in his sure hands and give it back to me unharmed.

Still, many people suffer from chronic pain that even the power of morphine cannot silence. Here, the pain-free Pakistanis give us hope: as they are healthy despite their missing protein, it should be possible to come up with pain killers that selectively inhibit that protein and thus break the vicious circle of pain.

But what about the torments of psychological pain? Our society is usually only willing to recognize and fight it when someone is obviously mentally ill. Many of us see the torments of drug withdrawal as self-inflicted and well-deserved. But even people who are not clinically depressed may find life so sad or threatening that they are unable to face it without drugs. How many of them are driven to suicide as the only way out? Today, it is 'immoral' and illegal to relieve the psychological pain of such unhappy people with 'hard' drugs. Is this the joyous resurrection of the now infamous anti-anesthetic movement? And could this mind-set be complicit in how, despite our enormous efforts, we are well on the way to losing the 'War on Drugs'? These questions are too complex for simple answers – and yet we have to look for answers. In this search, we will again have to fight both the complexity of our bodies and the power of our irrationality.

THE LITTLE BROTHER

THE LITTLE BROTHER

The diagnosis came late: for two thousand years, the corpses had lain buried in a cave near Jerusalem until scientists came along to show that almost half of them had harbored tuberculosis and leprosy bacteria. Similar results were found in an Egyptian sanctuary from the fourth century AD, a tenth-century Hungarian sepulcher, and a Swedish graveyard from the Viking era. Leprosy is one of the oldest known human diseases, and tuberculosis has usually been its brother-in-arms. But in fact, *Mycobacterium tuberculosis* and *Mycobacterium leprae* are not just brothers-in-arms; they are genuine brothers.

These two pathogens are members of the far-flung family of mycobacteria, peaceful soil dwellers that consume the remains of other organisms and thus assure the soil's fertility.

When humans began domesticating livestock about ten thousand years ago, some mycobacteria were able to make the evolutionary leap from the soil to cattle. Then it was not far to the herdsman – and as unhygienic cities developed, *Mycobacterium tuberculosis* became a scourge for humans and animals alike. At about the same time, another of these refugees from the soil specialized almost entirely in humans, adapted its genome and its way of life, and became *Mycobacterium leprae*, the agent of leprosy.

These two brothers are still similar in many ways. They look alike, multiply much more slowly than most other bacteria, and protect themselves with a layer of exotic lipids. Still, each has gone its own way. The tuberculosis pathogen became the big brother and took the path of merciless destruction, attacking the lungs and other vital organs of its victims. In the seventeenth and eighteenth centuries, it killed a quarter of the population of Europe and has probably been responsible for the death of more humans than any other pathogen. For a long time, the feared disease known as 'consumption' could only be fought with rest, a healthy diet, and a sunny, dry, and quiet environment – as in Thomas Mann's *Magic Mountain*. After World War Two, new kinds of chemical magic bullets made it possible for us to almost eradicate tuberculosis – but only in wealthy countries. Worldwide, the big brother is still lodged within one person out of three. In fact, an X-ray has shown that it settled in my own right lung when I was a child, where it is probably still lying in wait for its chance. Every year, it succeeds in making eight million people sick and killing two to three million of them. In the past few decades, it has also learned to combine forces with the AIDS virus, and a few of its variants in southern Africa have even learned to resist all of our chemical weapons. The dangerous advance of tuberculosis led the World Health Organization to declare a state of emergency regarding tuberculosis in 1993. The big brother is still far from defeat.

The leprosy agent also kills – but it prefers gradual attrition to rapid assault. It destroys the electrical insulation of skin nerves, disfigures the face with grotesque swellings, and finally makes fingers, toes, hands, and feet fall off and eyes go blind. Its victims become outcasts – the living dead. *Mycobacterium leprae* is the cruel little brother. When the two brothers still hunted us jointly, the little one mostly got short shrift, because the big one murdered its victims too quickly for the little brother to get its chance. Medical science has now forced the little brother to its knees with a recently developed three-drug cocktail; after one dose, leprosy victims are no longer infectious, and they can be cured in six to twelve months. When treatment begins early enough, even the disfiguring swellings disappear. But in the poverty-stricken parts of the world, the little brother still attacks several hundred thousand people every year. As nobody wants to take care of them, they hide their illness for fear of being outcast – and thus transmit it to others. Leprosy may not have learned how to resist our drugs yet, but it can still hide behind poverty and ignorance.

Why are the two brothers so different? Although they were the first bacteria identified as pathogens in the nineteenth century, the question remained unanswered for a long time. We did know a great deal about the tuberculosis bacterium, but we could not compare it carefully to the leprosy bacterium because the latter could not be grown in the laboratory, making its study difficult. Even today, *Mycobacterium leprae* stubbornly refuses to grow in artificial cell cultures; outside the human body, it reproduces only in mouse paws, in the cooler tissues of small rodents, and in some armadillos. And even there, it only divides once every two weeks – about one thousand times more slowly than most other bacteria. A few years ago, though, we deciphered the chemical structure of the complete genome of both brothers and can now finally explain why they are so different.

The big brother has carefully preserved the heritage of more than 4,000 genes passed down to it by its ancestors. With this genetic arsenal, it can adapt to quite varied environmental conditions, produce most of its own components, and come up with ways to foil our drugs. In contrast, over half of the little brother's genetic heritage has degenerated, so its genome is now riddled with over 2,400 mangled genes. This still leaves 1,600 intact genes in this genetic wasteland, but they are no longer even close to giving the little brother enough information to burn food, satisfy its own energy needs, and live independently. As a result, it can only reproduce inside human cells and the cells of a few animals. Its information deficit might also explain why it has not yet found a way to resist our medications.

Whoever wants to develop a strong character must be willing not only to learn new things, but also to put familiar things aside. This is not just true for us humans. According to the musician Pierre Boulez, the willingness to disregard tradition is a measure of a culture's strength. And the alteration or destruction of inherited genes is part of the development of all novel organisms. When the leprosy bacterium gave up half of its genetic inheritance, it took a great risk. But it was that very risk that endowed it with its unique biological character. Is the little brother less intelligent than the big brother – or more imaginative? I don't know. But for me, it is the more interesting of the two.

THE VOICE OF SILENCE

THE VOICE OF SILENCE

Silence has forsaken me. It slipped away a few years ago, leaving behind a soft chirping in my ears. In the soundless hours of the night, it brings back the sleepy summer fields of my childhood – and makes me wonder about the marvels of my sense of hearing.

My ears measure variations in air pressure, reporting them to my brain as electrical signals. None of my other senses works as fast. The eyes can distinguish twenty images per second at most – but ears react up to one thousand times faster, allowing us to enter the magical realm of sounds, from a violin's shimmering overtones at frequencies of around twenty thousand cycles per second all the way down to the deep bass of the organ at fifteen cycles per second. None of my other senses is as precise. I can distinguish sounds whose frequencies differ by less than 0.05 percent. And none of my other senses is as sensitive: my ears can respond to sound-induced vibrations that are smaller than the diameter of an atom. Working together, my two ears not only determine how loud a sound is, but also compare its time of arrival at each ear with almost uncanny precision. This allows them to tell me where it comes from and give me a spatial image of my surroundings no matter how little light there is. And still my hearing is nothing compared to that of an owl, whose lethally precise ears can locate a rustling mouse at a great distance even when it is pitch dark.

The organ that works these wonders is barely larger than a marble; it is kept safe and sound in my temporal bone. Its core is a spiral canal filled with fluid; above and below it are two elastic bands. About ten thousand sound-sensitive cells are lined up along the lower band like the steps of a spiral staircase. They have fine hairs on top, slightly resembling shaving brushes; the tips of the hairs touch the upper elastic band. A thin membrane separates this staircase from the middle ear and transfers air vibrations to the fluid and the two elastic bands, bending the tips of the hairs. Even the slightest displacement of these tips changes the electrical resistance of the hair cells; the resulting electrical signal races through the associated nerves and reaches the brain's auditory centers almost immediately. Each hair cell differs from all the others in the length of its hairs and the stiffness of its cell body. Things that are larger and more flexible vibrate more slowly, so each cell is tuned to a different pitch. Although the response ranges of the individual cells overlap, my ear and my brain take this overlap into account to give me a rich and differentiated sound palette.

Why does a hair cell in my ears react so much faster than a photoreceptor in the retinas of my eyes? When light reaches the retina, it first triggers a series of relatively slow chemical reactions that only later produce an electrical signal. In contrast, when sound displaces the hair of an auditory cell, it opens channels for electrically charged potassium and calcium atoms in the cell's membrane, immediately generating an electrical signal. Our eyes first have to light a fire under a steam engine, so to speak, which only then drives a dynamo to generate electricity, while our ears simply close the circuit of an already fully charged battery.

An auditory hair cell is extremely fragile. If the sound it registers is too loud, or too persistent, the cell dies and never grows back. In the evolution of the human species, ears apparently needed to be sensitive rather than robust; after all, extremely loud sounds (except for thunder, hurricanes, and waterfalls) are an 'accomplishment' of our technological civilization. Rock concerts, jet engines, discos, and jackhammers are inflicting hearing loss on more and more people, who then prefer their music quite loud and endanger other people's hearing as well. But even without excessive exposure to sound, our ears inevitably lose hair cells with age, above all those for higher frequencies. Like most older people, I no longer hear sounds with frequencies above eight thousand cycles per second. I can live with that, but such an impairment would end the career of a concert violinist, who must hear high-frequency overtones precisely in order to play higher notes in perfect pitch. And in general, the social and financial costs of hearing loss far outstrip those of blindness.

As is true for every signal, the quality of a sensory perception depends on the 'signal-to-noise ratio' – the relationship between the strength of the signal and any random background noise. A healthy ear can pick up noises over one million times weaker than the loudest sound we can tolerate. This impressive signal-to-noise ratio not only gives us a rich universe of sound, but also makes us virtuosos in deciphering even extremely complex acoustic signals. A high signal-to-noise ratio also allows for silence at the right time, turning silence itself into a signal. What would the two opening beats of Beethoven's Eroica symphony be without the subsequent silences? Aren't dramatic pauses before important statements the most striking feature of a masterful speech? And Wittgenstein's famous warning that 'what we cannot talk about we must pass over in silence' makes me wonder whether logic itself might demand the punctuation of well-placed silence.

Why does my hearing now deny me this silence? Are some of my auditory nerves sending phantom signals to my brain after the death of their hair cell partners? Or have the membrane channels for electrically charged atoms in my aging hair cells begun to leak?

The cells in my body work so well together because each of them only turns on the genes it needs for its particular tasks. My cells know a lot, but they only say what is necessary. In a typical cell of my body, then, most genes are silent. But now that my aging body is slowly losing control over them, they are getting restless. My skin spontaneously produces patches of brown pigment, and unseemly hairs sprout on my earlobes. If only a gene controlling my cells' growth does not raise its voice at the wrong time and in the wrong place to give me a diagnosis of 'cancer'! For genes, saying nothing is just as vital as saying the right thing. Even they know the virtue of silence.

CLOCKS OF LIFE

CLOCKS OF LIFE

After wintry Basel, the Australian sun was pure heaven. Yet I did not feel well, for my inner clock was still set to the cycle of day and night in Europe. This inner clock follows the earth's rotation; it is called my circadian clock because it roughly corresponds to one day (in Latin 'circa diem'). Other clocks tick in me as well, more slowly than the first, but even more inexorably. They once timed how I developed in my mother's womb, when I grew my first facial hairs as a young man, and when those hairs began to turn gray. Today, they also measure how much time I still have left. They are the clocks of my destiny – constantly running hourglasses no mortal has ever been able to turn over. In contrast, my circadian clock is an oscillator that regulates many of the cyclic processes in my body, as well as the overall rhythm

41

of my daily life. Each of these clocks is an intricate, self-controlled system – a wheel within a wheel within a wheel. The care life puts into providing double and triple backups for important regulatory mechanisms never fails to astonish me – and makes me fall silent in awe.

My body's circadian rhythm is a legacy from the dawn of life on earth. Over 3.5 billion years ago, some bacteria had already begun to harvest the energy of sunlight, and they must have quickly 'learned' to curb fast, energy-intensive processes at nightfall, such as chemical synthesis and growth. Their descendants today also turn down their metabolism at night; even if kept in constant darkness, they maintain this 24-hour rhythm for quite a while. In the wild, bacteria with dysfunctional circadian clocks use the energy of sunlight too inefficiently to have any chance at all against normal bacteria.

This bacterial clock is surprisingly simple. It is primarily made up of three proteins and ATP, a phosphorus compound present in all cells. These four molecules react with each other in a spontaneously oscillating system that can be reconstructed in a test tube with just these four players. In intact bacteria, this chemical clock is connected to the day-and-night cycle by a yellow, light-sensitive protein. How this connection works is not yet clear, but without it, the chemical clock falls out of step within a few days. A more recent, but also more complex light-regulated clock ticks in the body cells of flies, as well as in small transparent fishes in which light reaches every cell in the body. But when evolution began to create larger animals that are opaque to light, it moved the light sensors to special organs on the surface of the body. In many lower animals, the light-sensitive organ sits on the back of the head; we humans carry it in our retinas. It uses at least five different light sensors: four kinds of rhodopsin in the rods and cones, along with the photopigment melanopsin, which occupies its own layer in the retina and is not involved in normal sight. These five light sensors jointly send their signals to a dense nerve cluster, only two to three millimeters across, that is found in the brain's hypothalamus. Within this nerve cluster, a set of genes turn each other on and off at intervals of 24.4 hours, acting as a central circadian clock. This period is somewhat longer than a day, of course, but the light sensors in the retinas reset this clock every day according to the actual day-and-night cycle. This mechanism even works in most blind people, for melanopsin can fine-tune the central clock on its own. But anyone missing both eyes will have a biorhythm that is out of synch with the day-and-night cycle, a condition that often leads to sleep disorders and depression.

By regulating the release of hormones and other substances into the blood, the light-regulated central clock in the brain synchronizes the subsidiary circadian clocks of all the cells in the body with the earth's rotation. One of these synchronizing substances is the hormone melatonin, which also contributes to jetlag. The pineal gland synthesizes it from the amino acid tryptophan in a process that runs at night and shuts down during the day. Melatonin stimulates our craving for sleep, but it is still not clear precisely how. Other hormones in the body are also subject to a day-and-night rhythm, but with significant differences from individual to individual. We all know people who work best in the evening – and others who are already at full power bright and early. The circadian clock in our body cells even keeps ticking when we live in darkness for weeks or remove cells from our bodies and keep them growing in test tubes. In the latter case, we can even show that every cell passes its day-and-night rhythm on to its daughter cells. But without regular fine-tuning by light, the individual clocks in the cells sooner or later get out of step. This has far-reaching consequences, as each clock steers the chemical inner life of a cell in ways that are still poorly understood. By monitoring metabolic processes in individual cells, we have learned that every process has a characteristic phase and that the phases of different processes are often temporally staggered. For example, a growing yeast cell curbs its respiration when it duplicates its genetic material just before cell division. What are the advantages of such phase-shifted metabolic oscillations? Perhaps they isolate processes that could interfere with each other. The highly oxidizing by-products of respiration, for example, could harm the newly formed genetic material, which is extremely sensitive to oxidation. Membranes could also be used to keep metabolic pathways apart, but temporal separation is simpler – and more elegant. It also offers new approaches to cancer treatment, as the metabolic phases of cancer cells often do not overlap with those of normal cells. If a growth-inhibiting cancer drug is administered at a time when cancer cells – but not normal cells – are at the peak of their growth cycle, undesired side effects should be minimized. As most current cancer drugs are unfortunately also quite poisonous for normal cells, such 'chronotherapy' could significantly improve the efficacy of many cancer treatments.

And what about the much slower hourglasses that steer the longer phases of our lives? None of our body clocks are more elementary for us than the ones that determine how long we live. What clocks make a mouse live at most four years, a cat thirty-eight years, a human about one hundred and twenty years, and a bowhead whale more than two hundred years? We still know very little about these slow clocks, but at least one of them operates in each of our body cells: if human cells are grown in a test tube, they divide at most fifty to a hundred times – and then die. How are these divisions counted? One factor appears to be the gradual shortening of chromosomes at every cell division; it affects both ends of the chromosome strands and can only be repeated until the chromosomes can no longer fulfill their normal function. Cancer cells can somehow bypass this shortening step, which may explain why they can grow unchecked and overwhelm normal cells.

Life span also seems to be under the control of 'aging genes'. Several such genes have already been found in worms, flies, and mice. When they are systematically modified, the animals' life spans can be doubled or even tripled. Unfortunately, we have identified so many aging genes that we are not sure which of them are the decision makers – and which ones are just subordinates.

A third clock that determines our life span is the gradual destruction of our cells by oxygen. Life on earth originated at a time when the atmosphere contained no oxygen gas, so the first cells did not put themselves at risk by using oxygen-sensitive materials. But when a few of them began to harvest the energy of the sun about 3.5 billion years ago, they released oxygen gas from sea water, and so organisms had to quickly come up with ways to protect themselves from oxidative destruction. Unfortunately, they have not yet been completely successful. The lipids, proteins, and genes in our cells are very oxygen-sensitive, and they inevitably succumb to oxidation as time passes. Most of this oxidative damage emanates from the highly aggressive by-products of our cells' respiratory organelles – the mitochondria. This explains why organisms that respire intensively generally have shorter lives.

Molecular research into aging is now so successful that it will probably give us a pretty clear picture of the biochemical and genetic regulation of life expectancy within the next decade. Will we then be able to slow down our clocks of life, or even stop them altogether? In fact, wealthy countries have already begun to do so for their inhabitants: a Neanderthal human lived twenty years on average, people living in the Paleolithic era averaged around thirty-three years – and a Japanese of today can expect to live to eighty-one. Current global life expectancy is sixty-nine years. We owe this primarily to better hygiene and healthier nutrition, but antibiotics, mass immunization, and education have also played important roles.

But no reasonable person would make just prolonging life a primary goal. We should aim at maintaining the quality of life at as high a level as possible until the end of life. Our aging society is under enormous pressure to do so, for age-related diseases are becoming an almost unbearable financial burden, even for wealthy societies. The United States currently spends about fifty to eighty billion dollars a year just on the care of Alzheimer's patients. If the life expectancy of Americans increases as expected, this sum should rise to a trillion dollars by 2050. A very small but highly vociferous group of researchers studying aging, like the British scientist Aubrey de Grey, are extremely optimistic: they are convinced that a combination of chemical and genetic strategies will allow us not only to stop the aging process, but also to extend our life spans to hundreds or even thousands of years. The overwhelming majority of scientists, however, are suspicious of such predictions, remembering the famous American journalist Henry Louis Mencken's dictum: 'There is always an easy solution to every human problem – neat, plausible, and wrong.' Hardly any serious biologist believes that, any time soon, we will be able to extend our life spans several times over by simultaneously slowing all of the many clocks in our bodies that govern aging. Nevertheless, science policy should give serious research into aging much more priority than in the past, for such research is at least as important as research on infectious diseases or cancer. Perhaps a day will come when we will be able to drastically extend our lives. But who can say now whether this will ever be desirable? The generation of our grandchildren or great-grandchildren will have to confront this important question – and answer it.

There are so many clocks in our bodies – but which one makes us feel how the Viennese waltz owes its swing to a subtle delay of the second beat? What inner clock was Rainer Maria Rilke listening to when he wrote his *Sonnets to Orpheus*? And which timepiece told the conductor Wilhelm Furtwängler to produce the subtle tempo variations which made his adagios come to life? These clocks measure fragments of a second – as well as the pulse of hours. I would so love to know which of my hundred billion nerve cells give me these clocks! For it is these clocks that give my life meaning and joy.

MY WORLD

MY WORLD

Am I alone? Can I share with others the world I see and perceive? Or am I a prisoner of my senses and the poverty of my language? When our ancestors still hunted in groups, solitude meant danger. Today, we are tormented by the anxiety of loneliness. Anxiety is the fear of the unknown, which science should be able to help us overcome. And in fact, biology does teach me that I am firmly embedded within the rich web of life that covers our planet. But it also tells me that every human being tastes, smells, sees, and feels the world differently – which also offers me the comforting knowledge that my senses make me unique.

In 1931, when a sudden draft whirled up a powder from his laboratory bench, the American chemist Arthur Fox accidentally discovered that not all people have the same sense of taste: his colleague at a nearby workbench immediately noticed a bitter taste, while Fox did not. We now know that the ability to taste that powder as bitter is an inherited trait. Identifying bitterness is important for us humans: most plant poisons are bitter, and so we have at least 125 different sensors for bitterness. For most of them, though, it is still not known which bitter substances they detect. But we do know that people can have quite different bitterness sensors. Almost every West African, for example, can identify Fox's powder as bitter. West Africans are the genetically most varied of all human groups, with more variation from person to person than any other group. A small cohort of them settled in Northern Europe twenty-five to fifty thousand years ago, taking along only a fraction of the genetic variants found in West Africa. Since then, Northern Europeans and their offspring have had to get by with the limited taste repertoire of that particular group of African emigrants – and thus without the ability to identify some sources of bitterness.

We taste not only bitter, but also sweet, sour, and salty – as well as *umami*, the taste of sodium glutamate, which gives many Chinese dishes their special flavor. (The spicy tastes of the many kinds of peppers, though, are primarily identified through pain sensors.) The taste sensors for sour and salty have not been studied much, but those for sweet and *umami* are well understood. There are about half a dozen basic types, as well as many individual variations, so people perceive sweet flavors to varying degrees. Still, nobody is completely unable to taste sweet or *umami*. Evolution apparently prevented us from losing sensors for those two tastes, because we need them to identify foods with lots of calories and nitrogen. But cats do not eat any sweet carbohydrates, so they lost their sensors for sweet and cannot taste any sweet flavors at all.

Our taste sensors help us choose between foods, of course, but they might even dictate our addictions. The tendency to become addicted to alcohol or nicotine is partly heritable, so individual taste sensors might give some people a special craving for the taste of alcohol or cigarette smoke. If this is the case, we should be able to find medicines that dampen the sensors responsible for such preferences. Soon, genetic tests will almost certainly be able to identify those with a tendency for alcoholism or tobacco smoking, as well as those who might prefer particular drinks or foods. (Let's hope we will not abuse such knowledge.)

Smells also determine the 'taste' of our food – and smells are a mysterious, puzzling, and magical realm. Around 400 different sensors in the mucous membranes in our noses allow us to identify millions of scents. Using them and our taste sensors, we can distinguish over ten thousand different flavors. Mice and rats would not find this remarkable, for their highly sensitive noses are equipped with over a thousand different smell sensors. But smells do more than just contribute to the taste of food; they also have strong emotional effects because they are processed in an ancient part of the brain responsible for emotion and memory.

A substance can smell different to different people. A few years ago, a Swiss researcher had young men wear the same undershirt for several days and then asked young women which of the well-worn shirts smelled nicest to them. As had been expected, the answers were quite different, for there is famously no arguing about taste – or in this case, about smell. But there was one significant finding: women mostly preferred the smell of the men most different from them genetically. This famous 'T-shirt sniffing' experiment is still considered suspect by some researchers. But if further experiments confirm the results, this would mean that the individuality of our sense of smell helps maintain the genetic diversity of the human race.

As long as we are fully aware of the smells we perceive, we can defend ourselves against their effects. But what if we perceive some scents only unconsciously, so that they lead us to do things that are beyond our control? The animal kingdom abounds with such scents: pheromones. With her pheromones, a female moth can beckon a male of her species a kilometer away, a queen bee can prevent her subjects from raising a second queen, and a female mouse can tell an interested suitor her menstrual status. The sensors for pheromones resemble normal smell sensors, but they are mostly in a special region of the nose, and they send their information to a special part of the brain. Mice have up to 300 or more pheromone sensors – impressive proof of how much such imperious scents direct their lives.

We humans might well be unconscious slaves of pheromones, too. If women share the same room for a long time, they unwittingly synchronize their menstrual cycles, because they emit an odorless substance that triggers menstruation in other women. The chemical structure of this substance has not yet been definitively mapped, but it is probably similar to the primary female sex hormone. Similarly, men emit derivatives of the male sex hormone whose scent has a pleasantly relaxing effect on many – but not all – women.

The idea that scents can control me behind my back, as it were, unsettles me, but recent research offers me reassurance. Our distant ancestors probably had almost as many smell sensors as rats and mice do – that is, at least 900. In the course of our evolution over the last 3.2 million years, however, more than half of these sensors wasted away, and now their ruined genes drift silently in our genome from one generation to the next. Our pheromone receptors have fared even worse – or, in my view, better: their genes are almost all mangled beyond repair. Something similar happened to the genes for the proteins that transmit pheromone signals to my brain. Even if our noses are now much less sensitive than those of rats, we should be proud of our ruined smell and pheromone sensors. In order to become *Homo sapiens*, we not only had to learn new things, but also had to discard much of what we had inherited. On our long evolutionary voyage, we had the courage to trade the magical, but hazy realm of scents for the clear precision of our eyes.

Our eyes work with four different light sensors. The one located in the retinal rods is very light-sensitive, but cannot perceive color. In darkness, we rely only on it – and then all cats are gray. In bright light, we switch to three color sensors in our retinal cones: one for blue, one for green, and one for red. They are less light-sensitive, but they show us colors – every single one of them in about a hundred different shades. Our brain weighs the signals from the sensors against each other, allowing us to see about two million colors. Many animals, such as insects and birds, have up to five different color sensors and can see many more colors than we do, as well as ultraviolet and infrared wavelengths we are blind to. In contrast, the first mammals were night hunters who made do with only two color sensors. As a result, almost all mammals today can distinguish only around ten thousand different colors. The ancestors of today's great apes then acquired a third color sensor, which probably helped them distinguish ripe and unripe fruits against a background full of multicolored leaves. Thanks to this additional sensor, the world of great apes – and humans – is ablaze with millions of colors.

But about four percent of all humans can see far fewer colors; lacking either the green or the red sensor, they are 'color-blind'. In contrast, some women even have a fourth color sensor and can distinguish up to a hundred million colors. Such women may not always find this beneficial, as the colors in photographs or on television screens probably look unnatural or even wrong to them. But they might well be able to use subtle color changes in people's faces to identify liars, and they probably grasp the essence of colored diagrams quite a bit faster than other people do. There are an estimated one hundred million such 'superwomen' worldwide. Men can only envy them. But before marrying them, men should keep in mind that about half of their sons will have only two color sensors. Here, love makes men not only blind but even color-blind.

So all people see, smell, and taste in their own ways, and the same is true of our hearing and our sense of pain. Our senses show us a world that is uniquely ours and remains closed to others. Does art spring from the desire to escape the confines of this narrow world and give it general validity?

CHILDREN OF THE SUN

CHILDREN OF THE SUN

In the beginning there was light – a radiant explosion of energy, which created the universe roughly 14 billion years ago. As the universe expanded and cooled, light waves weakened into longer radio waves, plunging the universe into darkness. During the 30 million lightless years that followed, part of the radiation condensed into matter, matter into gas clouds, and gas clouds into galaxies. The immense gravitation within these galaxies forced atomic nuclei to fuse with each other and to release formidable amounts of energy as light. As the atomic fires of these first stars ignited, light returned to the universe.

Some stars found no stable orbits within their galaxy and, along with the remainder of the gas cloud, hurtled into a black hole in the galaxy's center. During their fatal plunge, these stars heated up to such a degree that they briefly outshone all the billions of stars of the galaxy put together. Most of these gigantic cosmic candles died roughly ten billion years ago. Yet we still see them, because the rapid expansion of the universe had propelled them – and their galaxies – so far into the outer reaches of the universe that their light is only reaching us now.

Many stars simply burned out, but some exploded, creating matter for new stars. One of these second-generation stars is our sun, whose nuclear fires ignited some 4.5 billion years ago. Like some other stars, it ejected matter as it formed, creating planets. On one of these, tiny clumps of matter evolved into ever more complex assemblies, which multiplied, moved and finally developed intelligence and consciousness.

I am a late scion within this aristocratic lineage of highly organized matter. This lineage is over 3.5 billion years old – plenty of reason to be proud of it. Very early on, my ancestors 'invented' the green solar collector chlorophyll, and in this way could feed themselves from light. To avoid the dangerous ultraviolet rays of bright sunlight, they developed sensors for short-wave blue light – and started to see the world in color. I carry at least five chemical offspring of this blue sensor in my retinas: three of them recognize either blue, green or red. As their color sensitivities overlap and my brain compares their signals with each other, I can see not only three, but millions of colors.

And yet I am close to blind, because I can see only a tiny selection of the electromagnetic waves that permeate the universe. These waves range from 1,000-kilometer-long radio waves to the gamma rays, just billionths of a billionth of a meter long, that are emitted by exploding stars. My eyes only sense wavelengths between 400 and 700 billionths of a meter, which my brain interprets as the colors of the rainbow – from blue-violet to deep red.

My sunlight-eating ancestors drastically changed the face of our planet. Feeding on an inexhaustible source of energy, they proliferated wildly, releasing oxygen gas from water as toxic waste. This gas gradually accumulated in the atmosphere, which until then had been oxygen-free. Soon some organisms 'learned' to use this dangerous waste to burn the remains of other cells – they 'invented' respiration. The food that sustains me is stored light energy – a faint reflection of the atomic fires in our sun.

Reflections of these fires even glimmer in the depths of our oceans. These are by far the largest natural habitat on our planet, yet below a depth of 1,000 meters they are dark, frigid and desolate, and often low in oxygen. How do living creatures orient themselves in this limitless darkness? How do they spot their prey – or their partners? And how do they recognize predators in time to flee? Much of this is still shrouded in mystery. Yet we do know that most deep-sea creatures accomplish these tasks with light signals. They usually generate them in special light organs, in which nerve impulses trigger chemical reactions between oxygen gas and substances in the body to create 'cold' light – as also happens in fireflies. Some deep-sea fishes even harbor light-producing bacteria in their lachrymal sacs, switching these inge-nious headlamps on and off by moving the skin around the eyes.

Ocean bacteria, too, may emit light. But unlike them, fishes, cephalopods and crusta-ceans usually emit their light in pulses which may carry information. The emitted light is nearly always blue-green, since such light pen-etrates water particularly well. For this reason, many deep-sea fishes make do with a single, blue-green sensitive visual pigment and trans-mit highly amplified signals to the brain. While they can see no colors, they can locate short or weak light pulses with high precision.

However, light pulses can also alert prey or attract predators. The black dragon fish, which lives at depths below 1,000 meters, avoids this danger by producing not only blue-green, but also deep red light, which is invisible to most other deep-sea animals. On this frequency, it can communicate privately with other mem-bers of its species and spot unsuspecting vic-tims. Its red light-emitting system first creates blue-green light that it converts to red light with the help of an additional pigment before finally 'purifying' the red light through a colored lens. To perceive the deep red light of its kin, the black dragon fish stores a variant of green chlorophyll in its eyes that effectively absorbs red light and somehow transmits its energy to the blue-green sensor of the retina. Since the fish cannot produce the chlorophyll itself, it probably ingests it with its food – but what this food may be is still a mystery. Whatever the answer, all light-generating substances as well as the oxygen gas they need for their reactions ultimately represent trapped solar energy. The pulsating points of light in the depths of our oceans are children of the sun.

Much about us and our world is mysterious and dark, and the darkness of our prejudices is more threatening than that of our oceans. Our reasoning gives us light in this darkness. Its glow is but weak – and yet it is the most wondrous of all the sun's children.

MYSTERIOUS SENSES

MYSTERIOUS SENSES

What is the world around me made of? I can see, smell, hear, touch, and taste it, but even though I use about one hundred different sensors for these sensations, as well as at least a tenth of my genes, they only open a tiny window on reality for me. My eyes see only a sliver of all electromagnetic waves, my ears are deaf to low and high tones, and my nose cannot smell millions of the scents that surround me. In order to escape the narrow confines of the senses, many people take refuge in esoterica, mysticism, or drugs. And in a way, science and art are also attempts to go beyond the limits of perception. What would the world look like if we could see ultraviolet or infrared light, hear ultrasound, and feel electric fields or the earth's magnetic field?

Many animals have such senses, none of which is more mysterious than the magnetic sense. Inside our planet, layers of liquid iron slosh around slowly, making the globe into a gigantic magnet. The resulting (and actually rather weak) magnetic field varies a great deal and switches poles about one to five times every million years. The last time it did so was 780,000 years ago, when our ancestor *Homo erectus* walked the earth. We can trace the alternation of the north and south magnetic poles back into the planet's early days because they aligned the magnetic particles in iron-rich sediments, which then remained fixed in their positions. The earth's magnetic field emanates from the South Pole, bends around the globe, and flows back into the North Pole. Its field lines run approximately parallel to the earth's surface at the equator and then slope more and more steeply down to the surface toward the poles.

Very early on, organisms began to orient themselves with these magnetic field lines. Thirty years ago, an American biologist noticed that some swamp bacteria always moved to the northern edge of a water drop, but then changed direction when he artificially switched the poles of the magnetic field around the drop. These 'magnetic bacteria' contain membrane-enclosed crystals of the magnetic iron oxide magnetite; long chains of these crystals act as compass needles and tell the bacteria's flagellar motor where to go. Dead magnetic bacteria still align themselves along the magnetic north-south axis, but no longer move towards the magnetic north. Magnetic bacteria grow best in layers of water that contain neither too little nor too much oxygen; they use their magnetic sense to figure out where the sea floor is and thus locate the appropriate depth. In southern waters, the same bacteria move towards the south instead of the north. Over the last two billion years, the magnetite crystals of extinct magnetic bacteria settled and solidified on the sea floor; along with non-biological layers of magnetite, the resulting sediments subtly distort the earth's overall magnetic field. As a result, the patterns of the magnetic field lines on our planet's surface are as characteristic as the outlines of the continents and oceans.

Many animals that undertake long, spectacular voyages in the course of their lives also orient themselves using the earth's magnetic field. Magnetic navigation is especially helpful for marine animals, which travel thousands of kilometers in the monotonous dim light of the seas. When a loggerhead sea turtle hatches from its egg on the beaches of eastern Florida, it immediately hurries to the east to the protective water, then swims with the Gulf Stream to the gyre of the Sargasso Sea and only returns to Florida years later. If it exits the gyre at the wrong place on its return trip and then strays into the cooler northern or warmer southern waters, it will usually die. The hatching young turtles first calibrate their magnetic sense with the light falling on the eastern beach; in the open sea, they probably measure the angle between the magnetic field lines and the sea floor. As this angle gets more and more acute toward the poles, it tells the swimming turtles their geographic latitude; how the animals take their longitudinal bearings is still a mystery. Mollusks, lobsters and many fishes also figure out their positions this way. All these organisms, as well as insects and mammals, use tiny magnetite crystals very similar to those of the magnetic bacteria. The rainbow trout has its ordered chains of crystals in special nerve cells around its nose. Each crystal is connected to the inside of the cell membrane by fine threads of proteins. Any magnetic variation exerts pressure on the membrane that opens mechanically sensitive membrane gates for electrically charged metal atoms. The resulting electrical signal is then sent to the brain to be decoded as information about the animal's position.

But the magnetic sense attains its greatest perfection in migratory birds, which home in on their far-away destinations with uncanny precision. How they manage this masterful performance is still largely unknown. Depending on environmental conditions, they probably identify their position with one or several different navigational systems, using the sun, the stars, and the earth's magnetic field. They seem to have two different magnetic sensors; every evening, they make rapid head movements that calibrate at least one of them with the setting sun. As in bacteria and fishes, one sensor uses magnetite crystals; it is found in the beak and measures the strength of the magnetic field. The other sensor is probably in the eye and measures the direction of the magnetic field lines. It seems to be made up of precisely ordered pigment molecules in the retina of the right eye. When struck by light, these molecules undergo a chemical reaction that either speeds up or slows down depending on the orientation of the earth's magnetic field. Quantum-mechanical calculations suggest that this field, although fairly weak, would be strong enough for this purpose. This synchronizes the circadian clock of higher animals to the day-and-night cycle. Birds thus exploit both magnetic information and light to identify their position. I would so like to know how a carrier pigeon experiences the earth's magnetic field!

Does it see it? And if so, does it see it as a color or as a pattern? It could also be that it feels, tastes, or smells it, depending on how the brain interprets the information sent to it by the bird's magnetic sensors.

Could the earth's magnetic field also influence me, without my being aware of it? My brain contains magnetite crystals similar to those of magnetic bacteria, fishes, and migratory birds, but there is no evidence that they give me a sixth sense. Magnetic healers are of a different opinion, of course; they promise that the magnetic rings, necklaces, amulets, and mattresses that they pitch have astonishing healing power. Apparently, many people are convinced, for the yearly turnover of such products comes to more than two billion dollars worldwide – even though there is no shred of evidence for their healing effect. On the contrary, everything suggests that these gadgets do not deliver what their apostles promise. Even Franz Anton Mesmer's highly regarded 'healings' by 'animal magnetism' in the eighteenth century were pure quackery that merely confirmed the ancient wisdom that faith can move mountains. If Molière had not died sixty years before Mesmer's birth, he might well have paid unflattering homage to him in one of his scathing comedies. Still, I cannot completely discount that even I might respond to the earth's magnetic field. For some people, changes in the magnetic field actually do seem to influence visual acuity and the sense of balance. It could well be that some of us have a hyperdeveloped magnetic sense and can thus perceive slight distortions in the earth's magnetic field caused by underground watercourses or ore deposits. The efficacy of divining rods may be just as unproven as that of magnetic amulets, but the former are at least scientifically more plausible.

I know so little about the world around me – and every question reveals to me the narrow limits of my innate senses. It often seems as if these senses were getting less and less important anyway, as they do not perceive the signals of our electronic world and cannot work fast enough, either. Electronic signals increasingly shape my daily life, but they often make me feel as if I lived in an alien world whose language I can neither hear nor understand. Today, isolated mountain peaks, deserts, seas, and forests are all immersed in electronic whispers from around the globe. Encoded military commands and time signals use radio waves that are kilometers long, and the daily news, discussions, and music travel on medium, short, and ultra-short waves. Television shows and instructions sent to airplane pilots, ship captains, and interplanetary probes use even shorter waves, along with the endless sea of digital signals from telephones, positioning satellites, and the omnipresent Internet. The signals carry petty chatter, tenderness, human tragedy, works of art, terror plans, medical discoveries, financial transactions, pictures of icy methane deserts on remote moons, and the designs of airports, computers, and entire cities. This modern communication system is the most gigantic and wonderful machine humans have ever produced – yet it speaks only to other machines and not to me. Its signals are not from my world. I may not be able to decode the whistling of a marmot, the chirping of a bird, or the scent of a flower either, but at least they make me feel part of a larger whole – even if I cannot feel the earth's magnetic field.

FATEFUL COLORS

FATEFUL COLORS

'That Schatz isn't so bad', the Graz school inspector growled over my head to my teacher, 'but that blond lad there in front, he'd be better'. Apparently, my brown hair made him think I was not 'Germanic' enough to recite a poem at a public birthday celebration for our (dark-haired) 'Führer' Adolf Hitler. More than six decades have since passed, and the mortification I felt at that moment has long since given way to an anger that keeps me from ever forgetting what deep wounds arbitrary interpretations of physical features can inflict.

Nothing shapes our first impression of a physically normal person as decisively as skin color. Groups of people differ in many other inherited characteristics, of course, such as their abilities to emit particular scents, to digest lactose, or to resist malaria. But skin color is obvious even from afar, and unlike body size or shape, it is common to most of the residents of a region. Since we unconsciously classify people with the same skin color as a unified group, this trait has influenced the course of human history as profoundly as epidemics, wars, and religions. Skin color is an ancient wellspring of the foul-smelling waters of racism; it even confused the thinkers of the Enlightenment into making rash judgments.

The color of our skin is the color of our fate. It comes primarily from a group of closely related pigments called melanins; they are produced by special skin cells that pass them on to the other skin cells and to our hair. Our irises also contain cells that produce melanin, but the cells keep their pigments for themselves. Melanins are mostly black to brown, but there are also lighter melanins, which are red or yellow. Dark hair contains a large amount of brown or black melanin, blond hair small amounts of brown melanin, and red hair almost nothing but lighter melanins. My 'gray' hair is a mixture of hair containing either a little bit of black melanin or none at all. I would prefer the brown of my youth, but now our loyal school inspector might finally accept me as a worthy member of the 'Germanic race'.

For animals, skin color is primarily a means of camouflage or of attracting a mate. For humans, though, it mainly serves as a shield against the ultraviolet rays of the sun – after all, we lost most of our protective body hair in the course of our evolution. Ultraviolet light not only damages our genes but also causes sunburn, dehydration, infections, and melanoma, a highly lethal form of skin cancer. The more the sun shines in a given region, the darker the skin of its inhabitants usually is. Dark skin possesses only two to three times as much dark melanin as light skin, but it is a hundred times less sensitive to sunburn – and almost five hundred times less prone to melanoma. No wonder the blazing sun in many parts of the world prevents the local inhabitants from losing their genetic disposition for dark skin. But twenty-five to fifty thousand years ago, African migrants settled in Northern Europe, where there is much less sunlight, and their dark skin became deadly: it kept the sunlight from helping their bodies produce the vitamin D that was essential to their lives, thus burdening them with infertility, brittle and bent bones, and an increased risk of infection. But Nature hurried to rescue the migrants by turning off their dispositions for dark skin and dark hair. This wide-ranging experiment, which would have been dangerous elsewhere, gave us humans a dazzling spectrum of blond, light-brown, and red hair.

A mouse controls its skin color with at least 125 genes. This is probably also true for humans, but until now we have studied only one 'color gene' in any detail; it determines the balance between dark and light melanin. Our body cells possess two copies of this color gene – one from our mother and one from our father. When both copies are intact, we produce more dark melanin than light melanin and have dark skin, dark hair, and dark eyes. When both copies are defective, we produce almost only light melanin and have light skin, light eyes, and red hair. And when only one copy is defective, our color type mainly depends on other genes and becomes difficult to predict. Defective variants of the color gene cropped up around the time of the settlement of Northern Europe. Today, about eight percent of the inhabitants of the British Isles overall and almost thirteen percent of the population of Scotland in particular have red hair. Why is this color type so widespread even though it offers no functional advantages? Could it be that it makes us attractive? Could light melanin be a sexual lure for humans? It is hard to overlook that Renaissance painters and the Pre-Raphaelite school had a weakness for red-haired models. Still, at least a few Neanderthals must have been red-haired and light-skinned, since defective forms of the color gene can also be found in the skeletons of representatives of that extinct human species. Could this explain why our *Homo sapiens* ancestors interbred with them? Who can fathom the intricate depths of love?

Melanin harbors many more secrets. Why is there dark melanin in our brain and in our middle ear? Why are melanin-free, blue-eyed cats mostly deaf? And why are red-haired women harder to anesthetize and more sensitive to pain than dark-haired women? The color gene and dark melanin may serve more purposes than we know about: not only is dark melanin the best sound absorber we know, it also protects the contents of cells from oxidation, as well as lower animals from invasive bacteria.

Fungi containing melanin thrive within the ruined nuclear reactor at Chernobyl, which has long since been shut down, yet is still highly radioactive. Their growth is sped up rather than slowed down by the radiation. Preliminary findings suggest their fungal dark melanin may turn the energy of the radiation into biologically useful energy. If this hypothesis is confirmed, then melanin would not be a fateful color for humans alone.

PORTRAIT OF A PROTEIN

PORTRAIT OF A PROTEIN

The New National Gallery in Berlin has a special treasure on the ground floor: Oscar Kokoschka's portrait of his friend and patron Adolph Loos. This expressionist masterpiece looks deep into the great architect's soul. Although neither the implied torso nor the dreamy expression reveal the combative innovator, the hands – painted larger than life and feverishly intertwined – give the painting a hypnotic power. They speak of despair and inner turmoil, yet they are still the resolute hands of a *Homo faber*, a builder of great things.

Pablo Picasso supposedly once said that art is the lie that reveals the truth ('El arte es una mentira que nos acerca a la verdad'). No work of art confirms this more clearly than Kokoschka's profound psychological portrait. It alters the model's outer form to reveal his inner essence. Looking at this painting, who could still believe that art seeks only beauty, while science seeks only truth?

Still, most of us would hesitate to consider Kokoschka's painting a scientific work. Our society sees art and science as separate, even antithetical worlds: art is intuitive, natural science objective; art seeks the particular in the general, natural science the general in the particular. From natural science, we expect the truth that reveals the lie.

We scientists have played our part in the development of this unhappy separation. If we have an artistic streak, we hide it behind wooden language and dry tables or charts. And when we do use images, we omit nothing, or we avoid putting too much emphasis on anything in them, for fear of being considered dishonest. However, this code of honor makes it increasingly difficult for us to describe complex systems. The metabolism of living cells, the earth's climate, and the structure of entire galaxies, for example, all give us too much information to be reproduced with our traditional thoroughness. Cramming all that information into a single image would render that image so complex as to obscure its object.

This is even true of individual molecules – such as the protein aquaporin, which my friend Andreas has been studying for many years. Aquaporin is a giant molecule made up of over nine thousand atoms; it is responsible for the intake and release of water through the membrane surrounding our cells. It is a complex of four identical protein chains, each of which spontaneously folds into a characteristic bundle and then combines with the other three bundles to form the functional aquaporin complex. To figure out how this four-part protein functions as a water channel, Andreas and his colleagues studied its three-dimensional structure. With years of hard work, they identified the position of every atom and described the complicated twisting of the four protein chains in the four bundles to within a billionth of a meter. Yet if they had shown me all this on a computer screen, I would only have stared at an incomprehensible tangle of points and lines

Portrait of Adolph Loos *(Der Wiener Baumeister Adolph Loos)* by
Oskar Kokoschka, 1909. © Foundation Oskar Kokoschka/2011
ProLitteris, Zürich, and bpk/Nationalgalerie, Staatliche Museen zu
Berlin (SMB)/Photo: Jörg P. Anders.

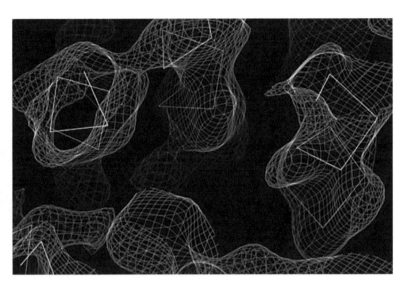

Portrait of the water-conducting protein aquaporin, viewed from the cell's surface. The yellow lines denote the contorted folds of the protein chain and the thin blue lines show the distribution of mass. © Professor Andreas Engel and Dr. Wanda Kukulski, Biozentrum, University of Basel.

that would not have told me anything I had not known already. A representation of every last detail of a complex object – be it a protein or a person – is bound to shroud the object's inner essence.

In an attempt to distill the character of their protein from its countless structural details, Andreas and his colleagues tried their hand at computer-assisted portraiture. Using pale screen colors for the less important sections of the protein chains, they emphasized important sections with sophisticated shading techniques and highlighted the repeating patterns of particular amino acids in the tangled chains with bright colors. They made some sections of the chains disappear altogether, thus giving more prominence to those that are crucial for channeling the flow of water. They also did not shy away from magnifying some of the details in the bundles to emphasize their precise form and their chemical features. They even gave free rein to their artistic flair with a colorful or structured background that made their scientific statement as aesthetic as possible. In the end, they created strikingly beautiful images that frequently adorned the covers of scientific journals.

But like all good portraitists, they were concerned less with beauty than with their model's inner life. As revealed in these portraits, aquaporin is a stout protein resembling an hourglass; it does not zoom around the cell's watery interior, but is instead firmly anchored within a membrane. To a chemist, the portrait makes it obvious why only water (and no other molecule) can pass through the constriction of the hourglass, and why a genetic modification of that constriction can disrupt water exchange through kidney membranes. And finally, the images also show that aquaporin is not very flexible, that it does not emit any biological signals, and that it is not an energy-driven pump but merely a passive channel for water. The aquaporin of these biochemical character studies is a solid pillar of society. In fact, the images suggest that it will have a long life in the cell. Like another famous portrait by Kokoschka, which uncannily anticipated the stroke suffered by the Swiss psychiatrist Auguste Forel, a protein portrait can even make predictions.

I have also come across portraits of proteins that determine whether a cell grows normally or runs rampant as a cancer cell, or of proteins with fearsome tentacles capable of tearing other proteins to bits. And then there are the portraits of aristocrats: proteins that act as tiny rotating turbines, delivering energy to our cells, and others, richly adorned with gorgeous pigments, that can turn the light of the sun into chemical energy.

Recognizing all of this in a protein portrait calls for the eye of a chemist or a molecular biologist; we can only see what we know. Anyone unfamiliar with proteins has to make do with the beauty of these images – as is also true of the paintings of Hieronymus Bosch or Max Beckmann, which can only be completely grasped by those who understand their rich symbolism.

Are portraits of proteins art? Many people would vigorously deny that they are – but with what arguments? Are these portraits less 'artistic' than highly detailed landscapes or still lives – or than the geometric op art abstractions of someone like Victor Vasarely? These are idle questions, for art cannot be forced into the templates of academic definitions. Still, the aquaporin portraits satisfy my longing both for the beautiful and the new – and thus show me the aesthetic dimension of natural science.

Molecular biologists want to understand life through its building blocks. The more complex these building blocks are, the more character they have. And in order to reveal their character, science is increasingly looking to its sister, art, for help. Science, too, must now lie in order to reveal the truth. The two sisters may remain separate, but they are now reaching their hands out to each other again.

TRACES

Where do we come from? What mysterious power created the highly ordered matter that makes me human? The search for answers to these questions gave birth to myths, but today we know that many of the answers are hidden in our genome. Every cell in my body has at least 25,000 genes; they are written in a chemical alphabet on giant, thread-like DNA molecules. All of my DNA threads put together constitute my 'genetic material' – my genome. If I could wander along its threads, I would find not only my own genes, but also about three million randomly scattered and mutilated virus genes. These genetic fossils make up about a tenth of my genome; they are evidence of the grim struggles our biological ancestors fought with viruses millions of years ago. These struggles reshuffled our ancestors' genetic material and may have helped turn them into human beings.

Viruses are not living organisms, but roving genes with a protective coat of proteins (and sometimes a fatty membrane, as well). As they have no metabolism of their own, they must invade living cells in order to reproduce. Some of them – the retroviruses – even smuggle their own genes into the host cell's genetic material. When an infected cell divides, it passes both its own genes and the viral genes on to its daughter cells. If the infected host is a multicellular organism reproducing sexually, the viral genes are not passed on to the next generation unless they have infected an egg or sperm cell. In such cases, the organism passes on the virus genes like its own, and they become fixed elements in its genome.

Retroviruses lying dormant in cells are ticking time bombs. They can suddenly leave the host's genome and become free viruses again, which break out of the host cell, invade other cells, and settle into the genetic material of their new host. One cycle of infection thus follows another. What leads dormant retroviruses to wake up again and set out on new conquests? We know next to nothing about that. But we do know a few things about how we defend ourselves against such invaders. Like medieval cities and castles, we rely on several rings of defense. In the outermost rings, we try to overpower the virus with our immune defenses or prevent it from attaching to our cells. If these defenses fail, we try to block the release of the virus genes from their protective coat or their insertion into our genetic material. When the retrovirus has broken through these defenses, too, patience is called for: as the genes of the virus, unlike our own, are not kept intact by constant natural selection, we have to wait until the ravages of natural mutations have crippled them beyond repair. For our ancestors and our species, this tactic has been largely successful: after about a million years, the smuggled-in virus genes turn into inactive genetic fossils that drift from generation to generation.

Our genome is thus not only a source of life but also a genetic ossuary. When we look through this ossuary with the tools of molecular biology, we peer deep into our own past and discern some of the forces which formed the genetic material of our ancestors. The battle between cells and retroviruses has been raging for hundreds of millions of years. So it is no surprise that there are so many virus fragments in the genomes of all mammals. The battle has not yet been decided: infectious retroviruses are still lying in wait in the genomes of almost all mammals, including our closest relatives, the chimpanzee. And since we became a species in our own right, more than a hundred different strains of retroviruses have managed to invade our egg or sperm cells and establish beachheads in the human genome. But we humans may be the first species to have won the battle against inherited retroviruses: almost all the retroviral genes we have been able to track down in our current genome are in all probability too mutilated to form infectious viruses again. Only in the case of one dormant retrovirus does it appear at all possible that it has preserved its infectious power in some humans, threatening them with disease.

But even genetic fossils that can no longer form infectious retroviruses do not always sleep peacefully. Some of them, despite being so mutilated as to be almost unrecognizable, still randomly jump around in our genetic material, leaving trails in the process. Like many of the aftereffects of wars, these trails are often harmful; they cause about 0.2 percent of all the mutations our genome suffers in the course of our lives. Such a mutation may damage a protein necessary for blood clotting, turning its victims into 'bleeders', for whom even the smallest wound can be life-threatening. But like all mutations, those caused by these restless virus fossils can sometimes be beneficial. A long time ago, a jumping virus fossil landed near a gene that stimulates the growth of human egg cells and thereby accidentally increased the fertility (and hence the biological fitness) of our species. Further, careful study of such genetic fossils has shown that the sudden development of mammals about 170 million years ago went hand in hand with a formidable wave of retro-viral attacks. Another such wave of invasions took place around six to seven million years ago, just before we humans bid farewell to our ape-like relatives. The upheavals of these biological wars probably forced the cells of our ancestors to retool their genomes in multiple ways in order to forge new weapons against the invaders. Is all this just a matter of temporal coincidence – or did these waves of infection trigger sudden evolutionary leaps? Could it be that this war, too, was the father of all things? Did it accelerate the development of our species – or even make our species possible in the first place? Do some of the mutilated virus genes in our body perform tasks we still know nothing about? And are such virus ruins only overpowered invaders – or an important part of who we are?

STARDUST

STARDUST

Anemia – the diagnosis did not surprise me. I had long attributed my constant fatigue to a bout of flu I had just recovered from, but I kept feeling more and more exhausted, then finally even dozed off during the sparkling overture of the *Barber of Seville*. I was enough of a biologist to suspect an iron deficiency, so I consulted my doctor. The diagnosis was a relief, for iron tablets would quickly help my body make hemoglobin again, and thus more red blood cells. In turn, those cells would give my body enough oxygen, so I would once more be able to enjoy life's splendors.

As reassuring as the diagnosis was, it was also absurd. Even though almost a third of our planet is pure iron, I had not managed to get my body its necessary daily iron ration of one to two thousandths of a gram. I may be separated from the earth's iron core by a stony crust, but even that crust is five percent iron. The brownish-red color of fields and deserts primarily stems from oxidized iron – rust. By weight, iron is the predominant chemical element on earth, yet iron deficiency is the most common human illness. It afflicts a third of the world's population, not only restricting the production of red blood cells, but also limiting the physical and mental development of hundreds of millions of children. Iron deficiency is a bloodsucker in the truest sense of the word, claiming more victims than such notorious killers as heart attacks, cancer, or AIDS.

The fault lies with the careless reptiles, who forsook the protective wetness of the seas for the land about 250 million years ago. With their tiny brains, they did not foresee that their new home would no longer overindulge them with dissolved iron salts. On the earth's surface, iron and many trace elements essential to life, such as copper and zinc, are mostly imprisoned in insoluble rocks and stones; they can be extracted from them by many bacteria and plants, but not by higher animals. As a result, pregnant mothers have to give the liver and bone marrow of their babies a supply of iron to start life with – and they often pay for their generosity with an undetected case of anemia.

Why have the land-based descendants of reptiles not learned to use iron from insoluble minerals? Perhaps they quickly learned that doing so would be dangerous, for iron is also poisonous for cells: its toxic dose is only six times higher than the daily requirement. A surplus of iron increases the oxidizing effect of oxygen gas and damages DNA and many other sensitive cell components. Free iron also encourages the growth of the dangerous bacteria that always lie in wait inside our bodies; just like us, they burn food with the help of iron-containing proteins. So our body has learned to regulate the intake and distribution of iron with the greatest care; above all, the iron in our blood has to be tightly sequestered in hemoglobin or other iron-binding proteins so as to withhold it from iron-hungry bacteria. When we come down with fever, the amount of iron circulating in our blood goes down even more, in a scorched-earth strategy aimed at starving bacterial invaders. Such a procedure mostly takes a toll on the defender, too – and the price I paid was apparently anemia.

Bacteria try every trick in the book to break through our iron blockade. Plague bacteria degrade iron-binding proteins and then make off with the iron. Other bacteria release iron-binding molecules of their own, and later pick them up again when they are saturated with iron. Still other bacteria meet their iron needs by stripping iron from the hemoglobin in aged red blood cells. But the gold (or is it the iron?) medal for inventiveness goes to *Borrelia burgdorferi*, the dreaded bacterial agent of the Lyme disease carried by ticks. This bacterium has managed to live entirely without iron by replacing all of it with other metals. A relentless battle for iron rages in us; in the course of human history, it has surely claimed more victims than all of the battles for silver and gold.

This battle for iron probably broke out quite soon after the origin of life on earth. According to some biologists, it even delayed the development of higher forms of life by a billion years. Originally, there was no oxygen gas on our planet, so the primal seas contained many reduced and dissolved iron salts that could serve the needs of living cells. But when some cells began to use the energy of sunlight, they released oxygen gas from the water at the surface of the sea. This gas oxidized sulfur-laden coastal rocks, causing them to release soluble sulfur compounds. As these washed into the oxygen-free depths of the seas, bacteria turned them into hydrogen sulfide. This foul-smelling gas caused the soluble iron salts to precipitate as insoluble sulfides, thus depleting the seas of dissolved iron. Simple organisms were still able to get by, but more highly developed cells had to fight for survival and thus stagnated. It took oxygen gas hundreds of millions of years to conquer the depths of the seas, where its oxidizing power blocked the formation of hydrogen sulfide. Only then did life enter a new iron age marked by the triumphant advance of complex, multicelled organisms.

In human history, too, the Iron Age was the beginning of a turbulent period of development. Until about four thousand years ago, metallic iron was a precious gift from the gods; it was only available when fiery meteorites occasionally fell from the sky. The resplendent tomb of the young pharaoh Tutankhamen, who died about 1320 BC, contained more gold than the country's royal bank, yet one of the apparently most precious burial objects was a small dagger with a golden hilt, a golden sheath – and an iron blade. Metallic iron did build up as a waste product in the smelting of copper, but it was much too difficult to process: its melting point of 1,537°C was long unreachable for the smelting ovens. It took many centuries until a few gifted smiths (apparently first in central Anatolia) intuitively learned how to work iron so that it became superior to bronze. Even then, iron processing long remained arduous and unreliable, making every good iron sword a precious work of art. The early Iron Age thus inspired legends of magic swords like Gram, Balmung, and Excalibur;

the myths of the Bronze Age tell no stories of wondrous weapons.

The iron on our planet and in my body is the dust of extinct stars. A billion years after the Big Bang, gigantic hydrogen clouds condensed into stars and galaxies and heated up so much that hydrogen atoms fused into helium atoms. This fusion released tremendous amounts of energy and thus lit the nuclear fires of the stars.

After hydrogen was used up, helium and its fusion products fused to form heavier and heavier elements, such as carbon, oxygen, and finally iron. Iron atoms are the most stable of atoms; their fusion uses up energy rather than generating it. Because of this 'iron curtain', many stars gradually burned out, releasing some of their mass into space – and thus some of their iron as well. But most of the iron on earth is the legacy of giant stars and special twin stars that suddenly imploded and, for a brief time, released as much energy as the billions and billions of stars of an entire galaxy put together. The cosmic inferno of such supernovas easily forged even the heaviest elements and then slung them far into the depths of the universe, where they combined with other cosmic matter to form new stars and planets – including our earth. Thanks to these well-travelled stellar cinders, our cells' iron-containing power plants can burn food and give us energy. The billions and billions of tiny fires that glimmer within us reflect the fires of long-extinguished stars from the depths of the universe. They remind us that we are links in a cosmic chain in which, for stars as well as for humans, every end is a beginning.

CREATIVE CRATERS

CREATIVE CRATERS

'Where do we come from?' This question has nagged us humans from time immemorial, but for a long time, only myths and holy books gave us any answers. About two centuries ago, however, biologists began to ponder the origin of life's diversity and concluded that organisms are not unique creations but ceaselessly change into new life forms. On this tree of life, we humans are only a tiny and late branch. But where are the roots of this tree? How did life on earth begin?

We will probably never answer this question definitively. But we do know that our earth already carried life quite soon after it had solidified. Before life appeared, the impact of an errant planet transformed the earth into a white-hot fireball and tore the moon out of it; in the hundreds of millions of years after that cataclysm, massive meteors left behind count-less craters, which have since eroded. But when things began to calm down again about 3.6–3.8 billion years ago, there was already life. Were those hot craters perhaps retorts in which inanimate matter turned into life? Could it be that the biblical Paradise was fatally similar to Hell?

The most primitive organisms still around today live in boiling hot geysers and sulfur springs, in fissures in the earth that are kilometers deep, and even in smoldering heaps of coal-mine waste. Their most extreme habitat, though, is around fissures in the sea floor. These fissures emit water at temperatures of up to 500°C; it is kept from boiling by the high pressure. When this water hits the ice-cold water at the sea floor, its cargo of dissolved metal salts precipitates as thick smoke, giving these undersea fissures the name 'black smokers'. This hot, lightless, and chemically highly reactive microcosm swarms with microorganisms that are the most primitive and robust of all known creatures. Some of them are even smaller than the wavelength of green light; others contain tungsten, which is otherwise extremely rare in living cells; many reproduce only at temperatures of around 100°C and stop growing below 80–90°C; others still survive temperatures of up to 130°C. Why their proteins are so heat-resistant, despite being otherwise quite similar to ours, remains a mystery. Under the microscope, these single-celled organisms may look like bacteria, but they actually have little in common with them. They form their own kingdom of life – the Archaea. Their genetic material reveals them to be the lowest branch on the tree of life. They are the closest surviving relatives of the unknown primordial organism from which all life on earth descends.

The metabolism of these Archaea still bears the marks of a primitive volcanic world. Many of them draw their energy neither from sunlight nor from the degradation of biomass, but from geochemical processes. Unlike most organisms alive today, they are not children of light, but creatures of the underworld. They have been found in a twenty-million-year-old underground source of hot water in the Mpo-neng gold mine in South Africa, one of the deepest mine shafts in the world. These inhabitants of Hades thrive on the reaction of hydrogen gas with sulfurous salts, which produces foul-smelling hydrogen sulfide. The hydrogen gas is formed when hot water hits iron-rich basalt. The life around us feeds on air and light – the life inside the earth on water and stone.

Although these subterranean microorganisms are apparently never short of energy, they grow billions of times more slowly than most other microorganisms. They are probably starved for biologically usable nitrogen, which limits life even at the earth's surface. How much life is there down in the depths of our earth – or deep inside other planets or moons in our solar system? Mars and Europa (a moon of Jupiter) may be forbiddingly cold, but they do have water: in warm subterranean niches on Mars, and on Europa even in subterranean oceans with their own sea floor. Both of these celestial bodies could offer subterranean habitats for Archaea-like microorganisms. Should we ever find life elsewhere in our solar system, it will probably resemble the life we have found in the depths of the earth's crust and in the cracks at the bottom of the sea.

We tend to forget what an incomplete and distorted image our senses give us of life on earth. More than half of the world's biomass consists of bacteria and Archaea, the majority of which we do not even know yet. We have identified fewer than ten thousand of them – less than a thousandth of the ten million species that probably exist. If just one of them had totally unexpected and novel features, it could revolutionize our current views on how life began.

Water samples collected from oceans all over the world have served as a powerful reminder of how little we know. In 2003, a team of American biologists modified a private yacht and sailed from Halifax down the east coast of North America, through the Panama Canal, and into the Pacific, where they passed the Galapagos Islands on their way to Polynesia.

On this two-year voyage, they took a water sample every 320 kilometers and examined the genetic material it contained – a quick, unambiguous way of identifying microorganisms without having to culture them. The result surprised even the researchers: every teaspoon of sea water teemed with millions of bacteria and at least ten to twenty times as many 'bacteriophages' – viruses infecting only bacteria. The expedition discovered countless new genes and bacterial species, even though these water samples came only from the surface of the sea. Who knows what the ocean's lightless depths conceal?

In a letter to the botanist Joseph Hooker, Charles Darwin surmised that life could have arisen in a 'warm little pond'. But he was modest enough to add that it 'is mere rubbish thinking at present of the origin of life; one might as well think of the origin of matter'. Since then, we have dared to do both, making breathtaking discoveries about the birth of the universe and the origin of our species. If Darwin's warm little pond was indeed the steaming water in a crater, then life had to adapt to the decreasing temperatures of the aging earth over the next few billion years. The question of where we come from is still waiting for an unequivocal answer, but for me, this is no reason to be upset. One of the fascinating aspects of life is that we still know so little about it.

MASTER CHEMIST
FROM THE DAWN OF LIFE

MASTER CHEMIST
FROM THE DAWN OF LIFE

I am one of the most complex forms of life on this planet – and a trained chemist to boot – but when it comes to chemical synthesis, most 'primitive' bacteria beat me hands down. Bacteria go back a long way; life's early struggles made them a hardened and inventive lot with many chemical tricks up their sleeve. My own cells are no longer as inventive, yet they have stubbornly held on to one of the wands with which bacteria perform their chemical magic. The story of this wand leads from the mountains of Silesia all the way back to the dawn of life.

Fables from the Alps have always been rife with such scary characters as the *Perchten*, the *Habergeiss*, and the *Krampus*. On winter nights, they were said to attack solitary wanderers, and deep under the mountains, *Kobolds* and *Nickels* made ores glitter so as to trick miners into thinking they had found veins of silver. But when smelted, these glittering ores produced not the precious metal, but only ugly slag – as well as *Hüttenrauch*, or 'smelter smoke', the murderous poison arsenic oxide. Only when the silver mines ran out did the Silesian miners find a way to profit from the 'Kobold ore' they had once scorned: it gave glass and glazes a deep-blue color. In 1737, the Swedish chemist Georg Brandt then showed that, alongside the already well-known elements arsenic and sulfur, this poisonous ore contained a previously unknown metallic element that he dubbed cobalt, which was later identified as element 27 in the periodic table.

The Silesian miners were not actually the first to make use of cobalt. The Egyptians anticipated them by at least three thousand years – and even they were only late epigones of the single-celled organisms that put the magic powers of cobalt to use in complex chemical reactions about three billion years ago. In order to link or separate carbon atoms in a chemical synthesis, these early organisms used the chemical power of cobalt atoms bound to particular proteins. Armed with these cobalt 'enzymes', they created new types of metabolic processes. They made some of these enzymes even more effective by surrounding the cobalt with an elaborate molecular cage, one of life's chemical masterpieces. It is a molecular spiderweb made up of more than one thousand atoms; at its center sits a cobalt atom, like a six-legged spider that holds tight with five legs while using the sixth to catalyze chemical reactions. This molecule was apparently created around 2.75–3 billion years ago when the primordial seas were still free of oxygen gas. More commonly known as vitamin B_{12}, it is dubbed cobalamin by chemists. For a long time, they tried in vain to elucidate its molecular structure; when the British biophysicist Dorothy Crowfoot Hodgkin finally did so, she used X-rays, not chemical methods. Then, in 1972, the Swiss chemist Albert Eschenmoser and the American chemist Robert B. Woodward succeeded in synthesizing the cobalt cage in the laboratory. It took more than a hundred chemists eleven years to achieve what is still considered one of the most difficult and virtuoso total syntheses ever accomplished.

In its earliest days, life experimented not only with cobalt, but also with nickel, iron, and manganese, for these metals were abundant in the primordial seas. The resulting metal-based enzymes enabled these early cells to perform unusual chemical reactions and to move into a wide range of new biological niches. But when a few organisms began to use sunlight and to free oxygen gas from sea water in the process, the oxygen caused sulfur-rich rocks to erode, washing much of their sulfur into the seas, where it made cobalt, nickel, manganese, and iron sink to the sea floor as insoluble sulfides. In their place, zinc and copper became the most common metals in sea water. The organisms alive back then were able to continue producing the metal enzymes they had already created, but they either replaced the metal in them with zinc or copper or developed completely new zinc or copper enzymes. The advent of oxygen gas triggered the development of higher life forms, but they invented few, if any new cobalt enzymes, making do instead with the ones they had inherited from their ancestors. Gradually, they even 'forgot' how to produce the cobalamin they needed to survive. They left that task to simple bacteria and now had to either eat them or live together with them in order not to perish. Today, only higher plants can get by without cobalamin. Algae, protozoa, animals, and humans must get small quantities of it in their food. The vitamin B_{12} that is essential to their survival is a legacy of the ancient 'cobalt age' from the dawn of life.

In order to survive in the long run, I only have to ingest one to two millionths of a gram of this vitamin every day. No other vitamin is effective in such small doses, perhaps because only two of the countless chemical reactions in my body require vitamin B_{12}. But without these two reactions, my cells would be oversensitive to oxygen and could produce neither enough energy nor genetic material for daughter cells. I get my vitamin B_{12} partly from my intestinal bacteria, but primarily from meat, milk, and eggs. In turn, the animals these products come from get their vitamin B_{12} from the bacteria that live in them and from plants contaminated with bacteria (or with animal excrement rich in vitamin B_{12}). Strict vegetarians who are careful about hygiene can thus run short of vitamin B_{12}. It can take a long time for that to become a problem, for the liver stockpiles enough vitamin B_{12} for many years, and the small intestine recovers a great deal of it before it is excreted. Vitamin B_{12} deficiency damages nerves and the brain and leads to the deadly illness 'pernicious anemia'. The cause of this disease is usually not an insufficient supply of the vitamin, but a failure of the cells of the stomach wall to produce a protein that enables absorption of the vitamin in the small intestine. We cannot cure this protein deficiency, but can effectively overcome it by injecting the vitamin into the muscles, thus guaranteeing a normal life for people suffering from this once deadly disease.

There is only one thousandth of a gram of cobalt in my body. It accounts for about as much of my weight as a human hair would on my car. But that hair is one of the countless threads connecting me to the web of life. I became a biochemist to understand the chemical events that take place in my body, and I had no idea that biochemistry would tell me about my distant ancestors and the breathtaking history of life. This history makes the history I learned in school (wars, coronations, the rise and fall of empires) seem small and insignificant. We usually begin our writing of history with the appearance of *Homo sapiens*, but now the molecular palimpsest of living matter has extended our temporal horizon by five orders of magnitude. Shouldn't historians take off their temporal blinders and look much further back into the past than they ever have before?

REFLECTIONS ON MYSELF

REFLECTIONS ON MYSELF

When Charles Darwin and Alfred Russel Wallace unmasked evolution as a game of chance, they shattered the long-standing belief that we humans are the crowning achievement of a long-term divine plan. Yet they also comforted us by implying that life on earth behaves as a single organism of which we are but a tiny part. Does this view of life still grant me freedom? Where are the boundaries that protect my individuality within the web of life? Nothing raises this question more urgently than the bacteria inhabiting my body.

When I was growing inside my mother, I was still clearly defined: every cell in my body carried my genome. But hardly had I left the protection of the womb when bacteria began to settle on me. Within a few weeks, they had colonized the surface of my skin as well as the mucous membranes of my nose, mouth, and digestive tract. Today, I consist of about ten trillion human cells and about ten to twenty times as many bacterial cells. Are these bacteria part of me – or only parasites? Where does my self end?

As bacteria are about a thousand times smaller than human cells, they make up only a small percentage of my body weight (about one or two kilograms). They are a variegated lot – my skin alone harbors up to five hundred species. And many of them apparently live only on me. They do not diminish, but enhance my molecular individuality.

Many of these bacteria are as crucial for my well-being as my own genes. They keep pathogenic bacteria at bay, and in the first few years of my life, they fostered the development of my immune system. When I was a starving child during the war, they provided me with folic acid, vitamin K, and vitamin B_{12}, for they can produce these essential nutrients from simple raw materials. My body cannot do this on its own, and our meager wartime fare was unable to provide my daily requirement of these nutrients. 'My' bacteria's chemical mastery may have then saved my life.

Not all my bacteria are peaceful, but as long as I am healthy and live reasonably, my immune system will keep them in check. When I eat unhealthy food, work too hard, or fight a virus infection, that defense fails, and bacteria can suddenly emit toxins, reproduce much more aggressively, and become acute threats to my health. Open wounds also destabilize the balance between me and my bacteria by giving them access to my blood and tissues, where they can run amok. And when I neglect to brush my teeth regularly, packs of oral bacteria cover them with a solid, impenetrable film and attack my dental enamel with the acids they excrete.

All known animals harbor bacteria, and many could not live without them. Particularly striking examples are some species of insects that can only live on the sap of particular trees. This sap is mostly a very one-sided diet: it lacks many amino acids that the insect needs to build proteins but cannot produce on its own. The bacteria living inside the insect, though, can produce them, which ensures the survival of their host. Many species of hosts have lived with their bacteria for millions of years and pass them on through their eggs to the next generation as carefully as they do their own genetic material.

No bacterium is as astonishingly proficient at living with its hosts as *Wolbachia*. Perhaps the most successful parasite on the planet, it lives in at least a quarter of all known insect species, as well as in many worms, crustaceans, and spiders. Although many *Wolbachia* hosts can survive without the parasite, they get several components from it that they cannot make themselves and that are not available in sufficient quantity in their food. *Wolbachia* probably infected one member of each of these species millions of years ago and has never left them since. The bacterium has been able to get away with this freeloading way of life by jettisoning about three quarters of its genetic material. This has not curtailed its propagation, for it is passed on through the eggs of infected mothers, and it changes the sexual behavior of its hosts to its own advantage. Depending on the host species, it can kill the male before the eggs hatch, turn a male into a female, or make males superfluous by getting females to give birth to infected daughters even without fertilization. In other cases, males infected by *Wolbachia* can only produce offspring with females who are also infected. The goal is always to infect as many female insects as possible and to give them advantages over males and uninfected females – all of which fosters the bacterium's own propagation. *Wolbachia* would make even Niccolò Machiavelli turn pale with envy. In the course of the next few million years, it may well transfer more and more of its genetic material to its host and turn into a cell organelle, in which only the trained eye of a molecular biologist can spot the telltale signs of a bacterial past.

I, too, harbor descendants of free bacteria that infected my distant ancestors 1.5 billion years ago and then settled permanently inside them. These bacteria had already learned to meet their energy needs by burning organic matter with the help of oxygen gas: they had invented cellular respiration. These respiring parasites supplied their host cells with the necessary energy to develop more complex forms of life. The host cells eventually took over more than 99 percent of the genetic material of their invaders, so that these now have only a few genes left. They have become an inherent part of me – my mitochondria. And the tiny remnant of their genetic material, which contains blueprints for only thirteen proteins, is now my mitochondrial DNA. It is my second DNA genome – much smaller than the one in my cells' nuclei, but no less important. If only a single one of its thirteen genes malfunctions or is lost, I am in deep trouble and may even die. Mitochondria can no longer live on their own or infect other cells; they get passed on by mothers through their eggs, just like *Wolbachia*. For mitochondria, males are a genetic dead end: I did not pass my mitochondria on to any of my three children.

About five to seven kilograms of my body are mitochondria. As they have been within me right from the beginning, I have always considered them part of my self. But now that I know about their origin, I am no longer so sure. And when I think about the one to two kilograms of bacteria that invaded me after my birth, the boundaries of my self begin to blur even further. Perhaps this is a good thing. Whoever takes his own self too seriously and timidly seals off its boundaries will distort his own perspective on the world's diversity and fall prey to primitive tribalism. This is as true for individuals as it is for peoples, nations, and cultures. Anyone who sees his own self as the measure of all things and man as the law-giving crown of creation must have slept through the last few centuries as soundly as anyone who still considers our earth the center of the universe. If modern biology now blurs the boundaries of my self, it does not diminish it, but gives it additional meaning and depth.

THE TWILIGHT OF IRON

THE TWILIGHT OF IRON

Two scientific revolutions have shaped my life: molecular biology and digital electronics. Now I am witnessing a third: the revolution of intelligent materials. For thousands of years, we have chosen our materials from what nature gives us in finished form. Today, we can also design new materials in the laboratory, manufacture them from pure chemicals, and endow them with information to fulfill specific tasks. This revolution makes even the highest grade of steel look like scrap metal.

Iron is still our most important material, even if aluminum, magnesium, titanium, glass, and ceramics are increasingly overtaking it. But, just like iron itself, they are only refined, transformed, or blended natural materials.

The first harbingers of the 'material revolution' were attempts to change the properties of natural materials by treating them with chemicals. In 1855, the Swiss chemist Georges Audemars soaked cellulose in a mixture of sulfuric and nitric acid and discovered that the resulting material could be dissolved in alcohol and ether and then spun into fine threads. Mixed with camphor, this became Celluloid, which is flexible and malleable when heated. Even though it is also highly flammable, it still provided an excellent flexible support for photographic emulsions, which made it the basis of the rapidly growing film industry. Similarly, formaldehyde transformed the protein from milk into a hard solid useful for buttons and buckles. But the actual revolution (and with it a new stage in our civilization) only began in 1909, when the Belgian Leo H. Baekeland used two pure chemicals to create a completely synthetic plastic resin: Bakelite. Its exact chemical structure was a mystery; it was only known that the two raw materials in it were linked into an irregular, three-dimensional network. Bakelite was an unattractive shade of brown; it was also brittle and unsuitable for producing fibers. Yet it was the beginning of the age of synthetic materials – and the beginning of the end of the Iron Age.

Two decades later, the German chemist Walter Bock invented the synthetic rubber Buna, and the American chemist Wallace Carothers came up with the synthetic fiber Nylon. Both materials are long chains of chemically pure building blocks that could be produced from only carbon, water, and air. Nylon is made up of long thread-like molecules whose two components are linked together head-to-tail about a hundred times. Shortly thereafter, the German chemist Paul Schlack went a step further with Perlon; it was similar to Nylon – but its chains contained only a single component. Nylon went on to become an icon of the postwar era: the flag Neil Armstrong raised on the moon in 1969 was made of it. After the successes of these synthetic fibers, the floodgates were opened. The flourishing petrochemical industry began to deliver a vast number of fully synthetic new materials and fibers with striking characteristics: polyester for comfortable clothes; transparent, shatterproof polyacrylate (Plexiglas); chemically inert Teflon; cheap poly-vinyl chloride for the construction industry; elastic polyurethane – and finally even organic materials that conduct electric current or gen-erate light more efficiently. As chemists under-stood how these properties depended on chem-ical structure, they could predict and change them almost at will.

These wondrous synthetics were the crown-ing achievements of a Golden Age of chemistry. In my youth, chemistry was a widely admired magician that gave us not only Nylon and Per-lon but also DDT, saccharine, brilliant dyes, and powerful new detergents and medicines. When I came down with diphtheria at age five, chemistry saved my life with the dye Prontosil, a precursor of today's sulfonamides. But all these achievements quickly became a matter of course and were then forgotten, for history does not know gratitude. A few decades ago, chemistry suddenly began to be despised as a polluter of the environment and a symbol of technological arrogance and contempt for nature. More and more people began to mistrust chemically synthesized medications, no matter how effective they were. And then chemistry had to cede its glamorous status to molecular biology. Chemistry, first a revered hero and then an evil sorcerer, ended up as a disenchanted simpleton.

Chemistry a simpleton? That may have been unfair, but it was still not conjured out of thin air. For all their sophistication, the new materials designed by chemists were primitive compared with living cells. The more we learned about the chemistry of life, the more clearly we realized just how unimaginably complex living cells are. The vast amount of information needed to construct a human cell is stored in our genes, the giant, thread-like molecules of DNA that carry the blueprints for at least 25,000 and perhaps even up to 100,000 different proteins. Like Nylon, every protein molecule is a long chain. But while the chains of a Nylon fiber are made up of only two components and run mostly straight and parallel to each other, protein chains have twenty different components and

fold up in the cell into complex, precisely defined coils and bundles. Thanks to their complicated structure, they can have quite varied
chemical characteristics and serve many different purposes. Proteins not only manage metabolism and communication between cells but
also provide structural support and serve as
raw materials (like the collagen in our tendons
and cartilage, or the keratin in our skin and our
hair). Despite all the differences between them,
the proteins in our bodies have one thing in
common: they contain much more information
than Nylon or stainless steel. Hence, they can
tackle much more challenging tasks, and can
even adapt to changed conditions. Their stabil-
ity is not static but dynamic.

Chemists have been inspired by living cells
and have begun to build highly complex materials that carry information needed to perform
specific tasks. One of the most impressive feats
in this field is the latest generation of bioactive
implants.

Fifty years ago, tissue-replacing implants
were developed by trial and error; our bodies

quickly recognized them as foreign and reject-ed them. Only as we learned more about the body's immune defenses and inflammatory processes could we rationally develop a first generation of implants that were chemically in-ert and physically as close as possible to the body part they were designed to replace. We still build them today (with special steel alloys, titanium, or plastics); they usually minimize acute immunological or inflammatory reac-tions, yet sooner or later the body rejects them, causing them to fail.

The next step was a second generation of implants that are either resorbed by the body or bioactive – but not both. What does 'bioactive' mean? In the past two decades, we have decod-ed many of the chemical signals that cells and tissue use to steer growth and healing. Most of this communication takes place between the cells' surfaces, which constantly emit protein molecules that attach themselves to other cells and give them orders: 'Stop growing immedi-ately; I'm here now', or 'Grow as fast as you can toward me so we can form a tissue together', or even 'Kill yourself; you're in my way'. Every tissue in our body is a beehive abuzz with countless messages. This is even true of our seemingly inert bone cells. In order to participate in this conversation between cells, proteins and other materials that steer cell growth are grafted onto bioactive implants to mimic the complex, information-rich structure of cell surfaces.

Second-generation implants include completely synthetic surgical threads that the body can break down into carbon dioxide and water, as well as titanium or ceramic implants whose surfaces stimulate the growth of bone cells. As cells respond not only to chemical messengers but also to the fine structure of surfaces, the tools of nanotechnology are used to engrave many such implants with tiny etchings one hundred to one thousand times smaller than a body cell. Nevertheless, about a third to a half of all skeletal prostheses of the first and second generations still fail after ten to twenty years. This rejection rate has not significantly diminished in the last few decades, in part because the clinical tests needed to develop new prostheses are quite labor-intensive and hence extremely expensive.

The third and latest generation of implants is not only bioactive but also resorbed by the body. These new implants are endowed with significantly more information and are thus more 'intelligent' than their precursors. They will first replace a tissue, then stimulate its healing, and finally just dissolve.

More and more often, the materials that make up living cells also inspire the development of innovative everyday materials. Mother of pearl, the iridescent coating on the inside of mussel shells, has alternating layers of thick, hard calcium carbonate and thin, elastic proteins that make it both hard and robust. The protein layers are only elastic when wet, so damp mother of pearl is twice as robust as it is when it is dry. The hard, extremely flexible spines of sea sponges are layered in a similar way, with the hard layers made of silica. Imitating the nanostructure of these biomaterials should make it possible to create artificial materials with unique and useful properties.

Still, science lives not on knowledge alone but also (and perhaps above all) on dreams. One such dream is to copy the supreme form of matter: a living cell. This would not only cause philosophical turmoil but also have great practical effects. Living organisms designed in the laboratory could capture sunlight much more efficiently than natural organisms, fertilize fields organically, degrade environmental poisons, or mine ore in inaccessible places.

How simple can a living cell be? Biologists recently discovered a bacterium that can make only 182 proteins. It is the simplest known form of life. With this limited endowment of proteins, it cannot produce many of its components on its own, so it has to live as a parasite inside insect cells. But a free-living bacterium probably needs only a few more proteins than that to survive – perhaps only 200 to 400. And we can already construct the necessary genetic material in the laboratory. Four years ago, the American molecular biologist Craig Venter used chemical robots to synthesize the complete genetic material of a virus that could infect and kill living cells. Recently, he and his group succeeded in producing something almost one hundred times larger: the genetic material of a simple bacterium, which they then used to replace the natural genetic material of a related bacterial species. By turning one bacterium into another, they created the first semi-synthetic form of life.

The next steps depend on what such an organism is to be used for. If it is supposed to degrade an environmental poison, the first step would be to choose a natural bacterium that is as simple as possible and that can already break down the poison. With chemical methods, the genes responsible for reproduction, basic metabolism and breaking down the poison could be boosted to maximal efficiency, and all the others could be removed. Finally, this custom-tailored, extremely reduced genetic material could replace the original genetic material. Ethicists would not find this problematic, but such a vision of 'artificial life' still makes people edgy. As has been true since the early days of biotechnology, strict rules will be needed in order to prevent unforeseeable accidents with these semisynthetic organisms (or are they materials?). Much later, we will probably even try to create fully synthetic single-celled organisms and give them as yet unimagined properties.

Living matter is the most complex matter we know. It is the result of almost four billion years of evolution, and it points the way to future materials. Would it be hubris to copy it for our purposes? Should we allow ourselves to trespass on worlds we previously shied away from as divine? And will life-like materials shape the high technology of our grandchildren? The revolution of intelligent materials has hardly begun, yet we are already building artificial materials carrying many times more information than anything we have ever created before. Why are we never satisfied? Perhaps matter endowed with information always hungers for more information. And wouldn't that be wonderful?

BEYOND GENES

BEYOND GENES

Who am I? How completely do my genes determine who I am – or could be? Am I unique – or only one of six billion identical bio-chemical machines? During the first half of my life, only great art could answer such questions for me. Philosophy and science let me down, as they had not yet realized – or had not wanted to admit – that the key to understanding life is its chemistry. During the second half of my life, physics and biology finally provided this insight and, after a long exile, reclaimed their position as cornerstones of philosophy. They revealed the awe-inspiring complexity of living matter, the uniqueness of every human being, and the common origin of all life on earth. Someday, they might also be able to show that we are more than just strictly programmed biochemical machines. If they do, they will free us of one of our most onerous philosophical slanders.

This slander is an unintended consequence of the scientific view of our world, and it has never been convincingly refuted. On the contrary, the discovery of genes, our growing understanding of how they work, and the deciphering of the chemical structure of our complete genome have even reinforced the idea that inherited genes rigorously determine our acts and our destinies.

But the wealth of information laid down in our genome might actually undermine the tyranny of genes. Living cells are the most complex form of matter that we know. The complexity of an object is a measure of the amount of information needed to completely describe it, so a living cell contains far more information than, say, a rock. In the genome of every cell, this information is stored in giant, thread-like DNA molecules in the form of a chemical alphabet. The genome of the simplest known bacterium, *Carsonella ruddii*, has 159,662 letters that describe 213 genes.

Most of them are blueprints for particular proteins, so *Carsonella ruddii* can produce 182 different proteins. That is not nearly enough for an independent existence, so this bacterium can only live as a parasite inside insect cells. Even there it may struggle just to survive, needing every single one of its 182 proteins. Thus, aside from rare mutations, all the cells of a *Carsonella* colony are essentially identical.

The genome that is stored in two copies in the nucleus of our cells contains 3.2 billion chemical letters. Although it is almost twenty thousand times the size of the *Carsonella ruddii* genome, it has only about 130 times as many genes – around 25,000. Our body cells carry a maternal and a paternal copy of almost every gene; theoretically, then, they can produce up to about 50,000 different proteins. In fact, they can produce many more, for they can read a gene in many different ways: they can read it from the beginning to the end, begin reading at different spots, or read only parts and then combine what they have read.

In this way, they can conjure up ten or more proteins from just one gene. And once they have finished making a protein, they can attach different chemical groups to it, altering its structure and function. As we cannot yet predict most of these changes from the structure of our genome, we do not know how many different proteins our body can actually produce: their number may well exceed 100,000. Our rich genetic heritage is thus a matter not only of the size of the genome, but also of the virtuosity with which it is put to use. Bacteria *read* their genomes; we *interpret* ours. We are like musicians in the seventeenth and eighteenth centuries, who could make any given basso continuo sound quite different by approaching it in a different way. Further, our brain cells seem to chemically change some of their proteins in response to environmental stimuli, so the variations of our cell proteins are practically boundless. This is even true of genetically identical twins: an identical twin brother of Roger Federer would look almost the same as his famous sibling, yet might well be a mediocre tennis player. The wealth of information laid down in our genome gives each of us the gift of uniqueness.

The information content of the genome and the freedom to interpret that information determine the rank of the organism in the hierarchy of life. An inflexible genome without much information is the archenemy of biological freedom and individuality. The more information a genome carries, and the more freely it can be interpreted, the more room there is for each individual organism to establish its uniqueness.

The ten trillion tightly networked cells in our bodies contain so much information that it may well be impossible to steer or predict the actions of a human being with any precision. Perhaps we have to come up with completely new kinds of thinking to understand systems of such complexity. Our natural laws, which derive from the experience of our senses, are only valid within certain limits: they fail at extremely small dimensions, with single atoms or molecules, or at extremely high speeds. Could it be that extremely complex systems also obey their own rules or laws?

Our genome is not so much a strict code of laws as a collection of flexible rules and instructions. The genes of our immune system spontaneously swap parts in order to give us as wide a range of protective immune proteins as possible. In the maturing brain of a mouse, short pieces of genes spontaneously trade places in the genome, which might well influence the development of nerve cells. Even in bacteria, pieces of genes can jump around in the genome when heat or toxins threaten the cells. That is, the environment talks to and can change our genes. Is this interplay precisely steered – or is it a game of chance? And if even the environment can play games with our genome, perhaps we do, too, without knowing it.

The physicist Erwin Schrödinger was the first to suspect that the hierarchical structure of living matter could impose the randomness of individual molecular reactions on an entire organism, making it unpredictable. In a typical crystal, all molecules inside the crystal are equal, but in the rigorous hierarchy of living cells, they are not: they each have their own rank. Some of them are present in such small numbers that their chemical reactions within the cell no longer obey the statistical laws of conventional chemistry, but are random and unpredictable. Cells have found ways to amplify such random fluctuations, transforming them into irreversible states. By imposing an element of chance, organisms can thus partly liberate themselves from the yoke of rigorous genetic programming.

Can this liberation grant us freedom of the will? The question remains open. We still know too little about our brain and our consciousness to understand what freedom of the will might mean.

But random fluctuations in the reactions of biological steering molecules readily explain why genetically identical nematodes raised in identical conditions react to heat differently and live for different lengths of time; why cells in a genetically homogenous bacterial colony respond to toxins or nutrients in distinct individual ways, and why genetically identical bacteriophages can infect their victims in several different ways. In its quest for diversity, life clearly does everything it can to fight the tyranny of genes. What parts of my existence are highly amplified molecular noise? How much does such random noise undermine my genetic programming? To some, this noise may reveal the divine breath, but to me it is a reminder of my miraculous existence as highly complex matter in our chemically primitive universe.

THE FOUNDATION
OF THINGS

THE FOUNDATION
OF THINGS

On November 4, 1856, the French sugar and alcohol magnate Louis-Emmanuel Bigo-Tilloy asked a young chemist at the University of Lille for help. Every year, Bigo-Tilloy's factory kept suffering great losses because the sugar in fermenting sugar-beet juice sometimes turned into acid instead of alcohol. Wouldn't it be possible to add chemicals to prevent abnormal fermentation? The young chemist promised to look into the problem. His name was Louis Pasteur – and the meeting of the two marked the birth of modern medicine.

At the time, Pasteur was only 34 years old, but he was already Dean of Natural Sciences in Lille and one of the most famous chemists in France. His instincts as a researcher were piqued by Bigo-Tilloy's problem, so instead of just running tests with a few chemicals, he set out to identify the cause of the problem. His microscope revealed that beet-juice samples that had undergone normal or abnormal fermentation harbored different microorganisms: normal fermentation starters contained round yeast cells, while abnormal starters contained oblong organisms. Clearly, it was the latter that made the beet-juice sugar turn into acid instead of the desired alcohol. Pasteur concluded that alcoholic fermentation was the work of living yeast cells. He then showed that adding these cells to beet juice ensured normal fermentation. Pasteur decided to continue to study these tiny organisms. He found that they did not come from dead matter like dust or air (as was commonly assumed at the time) but always descended from identical ancestors, just like animals, plants, and humans. The beet-juice 'disease' was thus nothing but an infection with the oblong microorganisms.

Next, Pasteur not only showed that other microorganisms turned milk or wine sour, but also discovered that those microorganisms could be killed by careful heating, which prevented acetification. Today, every child knows about 'pasteurization'. Pasteur's research culminated in the discovery that even many human diseases are infections of the body by microorganisms. This was one of the most important medical discoveries of all time. It quickly became possible to use chemical disinfection to prevent or at least check many contagious diseases. At first, though, Pasteur's ideas were scorned by many influential doctors, who found it absurd that such tiny organisms could make us sick or even kill us. In 1860, one of them gave free rein to his disdain for such nonsense in *La Presse*: 'I am afraid that the experiments you quote, Mr. Pasteur, will turn against you. The world into which you wish to take us is really too fantastic.'

A few years later, Pasteur and other scientists isolated weakened forms of pathogenic bacteria from animals and humans. Such bacteria were no longer a danger to the human body, but they could still trigger its natural immune defenses. People 'immunized' with these weakened bacteria were also protected against the pathogenic form. From that moment on, vaccination began to transform medicine – and the world.

Louis Pasteur, the chemist, became one of the greatest doctors of all time. The foundation of his genius was his insight that an illness can only really be cured when its causes are known. Before him, medicine was a world of pseudo-cures: baths, bleedings, mercury fumigations, laxatives, and costly diets. Such approaches were almost always ineffective or even harmful, for they were not grounded in the knowledge necessary to produce genuine cures. Pasteur's approach was not only effective, but often simpler and less expensive to boot.

Toward the end of the Second World War, medicine went through a crisis. The discoveries of Pasteur and his colleagues and the advent of antibiotics had significantly checked bacterial diseases, but viral diseases were claiming more and more victims. Viruses are much simpler than bacteria, and mostly much smaller, too. They are not living organisms but essentially wandering genetic material that can only reproduce in living cells. When they infect us, they become part of our own cells, so most of our antibiotics cannot touch them. One of the most feared viral diseases of that time was poliomyelitis; every year, it made innumerable children into lifelong invalids. Could medicine build on its earlier successes, which were all still a matter of living memory? Just like Bigo-Tilloy a century earlier, the public demanded rapid, goal-oriented remedies: more doctors, more iron lungs, more rehabilitation centers. Wealthy countries could fulfill such wishes for their citizens, but the costs were enormous, and the successes not even modest. This should not be a surprise, for these were just pseudo-cures: medicine lacked the basic knowledge to prevent polio effectively. Researchers from a wide range of disciplines first had to produce the necessary knowledge – and often they did not even know they were laying the groundwork for victory over polio. In the early 1950s, pure strains of the polio virus were finally grown in the laboratory, and three of them were shown to cause polio in humans. Armed with this knowledge, Jonas Edward Salk and Albert Bruce Sabin were then able to develop effective vaccines that practically eradicated the disease in Europe and the United States. A genuine treatment had been found – again, simple, effective, and cheap.

The discovery of antibacterial sulfonamides and the first antibiotics may have been the result of an accident and the good nose of a few researchers. But for those weapons to be kept sharp, years of work had to be put in to find out how they worked. Only thanks to this knowledge can we now systematically adapt such medicines to renew their effectiveness against bacteria that have developed resistance. In the fight against bacteria, we may win one battle after another, but we will never definitively win the war, for billions upon billions of quickly growing bacteria will always find new ways to overcome our defenses. In addition, virus strains still unknown a few decades ago (such as the AIDS virus) are becoming more and more dangerous for our species.

These days, medicine is increasingly fighting diseases that are caused not by external threats but by changes in our own body cells: arteriosclerosis, diabetes, schizophrenia, most forms of cancer, and such inflammatory and degenerative diseases as kidney failure and arthrosis. Hardly any of these feared diseases can yet be effectively prevented or cured – we still know too little about what causes them. As in the past, this ignorance forces us to turn to pseudo-cures. Doctors today have every right to be proud of their brilliant achievements. They perform open-heart surgery, remove deep-seated tumors, implant artificial joints, or transplant entire organs from one person to another. But all these methods fight only symptoms and not real causes; they are powerless to prevent arteriosclerosis, tumors, diabetes, or heart or kidney failure. And on top of that, they require enormous technological investments that generally make them much too expensive to have a major impact around the world.

Unsurprisingly, then, the public is getting impatient and calling for targeted, 'relevant' research that focuses on specific diseases. This call was first heard about forty years ago in the United States; it has since echoed through most western industrial countries. Neither scientists nor politicians can take this call lightly, for it is understandable. Although global investment in basic medical research was relatively modest until a few decades ago, biology and medicine still scored one triumph after another. In fact, all of Pasteur's research into contagious diseases probably cost the French government less than the yearly budget of a modern cancer research institute.

Today, biomedical research is the engine of the mighty pharmaceutical industry, devouring ever greater sums. In many fields, research itself has become big business. When a taxpayer sees imposing research institutes popping up everywhere and learns from the media that such institutes have budgets of millions and millions, sooner or later he will wonder what they have achieved. Don't the doctors already know enough to finally cure such diseases? Isn't it just a matter of being more systematic about putting that knowledge into practice, as the United States did with their atom-bomb and moon-landing programs? Shouldn't researchers be ordered to work directly on curing cancer instead of wasting time and money studying exotic-sounding problems?

Sometimes, I imagine answering such questions myself. Perhaps I could then spare the scientific world a few difficulties and help reduce the general unease about basic research. I would tell people that human cells are at least a thousand times more complex than bacteria, which means that they require very delicate, specific, and complex research methods. We are far from knowing enough about the biochemical processes in healthy body cells, let alone why they sometimes go awry and make us sick. Impatience, I would add, is one of science's archenemies, and good scientific research is as patient as true love. So it would be counterproductive to limit research to specific diseases. Such narrowly focused research can only succeed if the theoretical knowledge for reaching the stated goal is already available. Only then can a massive deployment of resources lead to quick success. When the United States decided to build a nuclear bomb in a very short time, or to send a man to the moon, most of the necessary basic knowledge was already available. For the cure of most organic diseases, though, this is not yet the case.

I would go on to reassure my fellow citizens: as even pseudo-cures can temporarily alleviate a problem, we are morally obliged to use them as long as we have not yet found any genuine solutions. We must completely exhaust the diagnostic and surgical potential of modern medical technology and support its further development as much as we can. But such work should not take the place of the search for genuine solutions. If we flag in this quest and limit basic research, we will only reap failure and disappointment in the end.

In democratic societies, it is always difficult for politicians to resist voters' calls for quick, partial solutions. This is true not only for medicine but also for many other problems in our society, such as the dangerous increase in crime in our big cities. Here, too, a genuine solution will surely only come from a fundamental understanding of the problem and its causes. In the long run, alleviation of tensions between ethnic groups and social classes, reduction of unemployment, improved education, and successful integration of migrants will be more effective than more police officers, better street lighting, and longer prison sentences. And yet such popular pseudo-solutions are precisely what the public clamors for. This is the old call for more bleedings, more iron lungs, and more organ transplants. Pasteur's example shows us that we will only solve the multi-layered problems of our society when we probe for the foundation of things.

VOICES OF THE NIGHT

VOICES OF THE NIGHT

For seven years now, I have been a professor emeritus – a retired professor. If I no longer have a laboratory, staff, or research funds, I also no longer have to fill out pointless forms, write reports for the desk drawer, and fight off sleep in unnecessary meetings. Yet my freedom still makes me giddy. It makes every day an experiment that can bring unexpected things to light – about science, about my former profession, or about me.

As a young biochemist, I usually worked in the laboratory until late at night – and sometimes even into the early morning. The quiet of the nighttime laboratory gave my thoughts free rein and helped me come up with new ideas. Today, when I lie awake at night and follow the trail of my memories, I miss that silence, for the voices of the night start troubling me with their questions. The voices are relentless, and I cannot lie to them. I try to resist them, but they come at the *Hour of the Wolf*, when my defenses are down and my thoughts drift in the no man's land between waking and dreaming.

Again and again, the voices ask what science gave me. It is not easy to answer that question, for it has so many answers. I wanted to become a scientist in order to learn what the world around me is made of. But I soon realized that scientific truth can very quickly prove to be false. One of my colleagues admitted this in a speech at a medical school graduation ceremony: 'We have done our best to teach you the most up-to-date scientific knowledge – but about half of what we taught you is probably wrong. Unfortunately, I cannot tell you which half.' Science does not reveal definitive truths but a way to approach truth. It is not about collecting and organizing facts but about believing that we can understand the world through observation, experiments, and reflection. Science also made me modest by showing me the narrow limits of human reason. Arrogance, hierarchy, and power have always been its archenemies.

Yet science never quite satisfied me. Shakespeare's sonnets, the Adagio of Mahler's Tenth Symphony, or Cézanne's visions of Mont Sainte-Victoire took me to an enchanted land beyond all science. This land let me see the world from a different angle, giving my view perspective and a sense of depth.

Again and again, the voices ask the question I fear most: 'Were you a good scientist?' All too often, I was not, for I was not always passionate, brave, and patient enough. Scientific success does not come from intelligence and originality alone: passion, courage, and patience are the most important qualities of a good scientist. Without them, you cannot call generally accepted ideas and dogmas into question and solve difficult scientific problems. Nor can you keep pursuing a goal for many years despite failures and setbacks. Science's main weapon, a hunger for knowledge, is blunt without the edge of intelligence. Yet even the keenest intelligence is powerless without passion and courage – which in turn are smoke and mirrors without the sustaining power of patience.

And the voices keep asking questions: 'Did you help your students and postdocs to be passionate, courageous, and patient?' Here, the answer hurts: 'Surely not enough.' I do not believe that passion can be taught, but courage and patience grow through interaction with courageous and patient people. Personal role models are paramount in the development of young people, so I tried my best to be a model of courage and patience for my research group. Mentors are a university's most important gift to its students. Why do I realize this only now?

And then comes another: 'What would you do better if you could start over again?' Here the answer is easy: 'I would take teaching at least as seriously as research.' By 'teaching', I do not mean the listing of scientific facts, but passing on my scientific experience and my personal understanding of science, the world, and existence. We professors should not just pass on the facts but also encourage young people to think independently, to liberate themselves from prejudice, and to seek answers to the great questions – questions about our existence and the essence of the material and intellectual worlds. All young people are looking for answers to such questions, even if they do not realize it, or do not want to admit it. When our educational institutions let them down, they will look for answers from gurus or religious fanatics. How else can it be explained that so many disciples of the infamous Bhagwan Shree Rajneesh sect had studied at the best universities in the United States? In retrospect, I can hardly believe that hundreds of gifted young people listened to my lectures for over an hour. What a singular opportunity I had to form them! Only too rarely did I take advantage of it, because I wanted to get back to my laboratory.

'How did science surprise you?', a voice asks. Here, too, I don't have to search for the answer very long: 'I expected solitary research and had no inkling of how much the community of other scientists would shape and enrich my life.' Great scientific discoveries are the children of solitude, yet they are never born in isolation. We scientists are working on a cathedral whose completion none of us will ever see. The community of our work gives us the strength to keep on going.

My nighttime visitors want to know so many things. Only at dawn do they fall silent. In order to better defend myself against their questions, I am now writing down my answers by the light of day. They are experiments – *essais*. Michel Eyquem de Montaigne saw his *essais* as experiments on himself – but maybe he, too, just wanted to appease the voices of the night.

ACKNOWLEDGMENTS

Most of these essays originally appeared between 2006 and 2008 in the *Feuilleton* section of the *Neue Zürcher Zeitung* and were collectively published as the book *Jenseits der Gene* in 2008 by NZZ Libro. Friends and colleagues from around the world watched over their genesis; their advice kept me on the path of scientific precision. My thanks go to

Maik Behrens
Urs Boschung
Daniel Burgarth
Christoph Dehio
Daniel Demellier
Andreas Engel
Albert Eschenmoser
Mariacarla Gadebusch
Bruno Gottstein
Claus Kopp
Walter Kutschera

Fritz Paltauf
Jonathan Rees
Daniel Scheidegger
Ueli Schibler
Anna Seelig
Joachim Seelig
Karl Stetter
Andreas Tammann
Michael Thomm
Rüdiger Wehner
Ingomar Weiler
Uwe Justus Wenzel
Roswitha Wiltschko

I also owe many valuable suggestions to my wife Merete, our children Isabella, Peer, and Kamilla, my brother Helmut and to Alexandra Stölzle. Special thanks go to my childhood friend Heimo Brunetti, whose breadth of knowledge and subtle feel for language contributed to almost every essay in this book. Even after translation, this book bears his unmistakable imprint; I dedicate it to him. This English edition owes its existence to Professor Wolfgang Schürer, who originally suggested it and, through his continued interest and generous support, made it possible. Publication was also aided by a generous financial contribution from the UBS Culture Foundation.

The translator's art is never an easy one, particularly if the text demands both scientific and linguistic precision. I am greatly indebted to Andrew Shields for his superb translation, which often added depth and luster to my German original. The translation of *Children of the Sun* was adapted from an earlier one by John Lambert.

ABOUT THE AUTHOR

Gottfried Schatz was born on August 18, 1936, in Strem, a little Austrian village near the Hungarian border. He grew up in Graz, but spent the year 1952 as an American Field Service-sponsored high-school student in Rochester, NY. After receiving his PhD in Chemistry from the University of Graz in 1961, he joined Hans Tuppy at the Biochemistry Department of the University of Vienna where he began to study how cells build their organelles of respiration – the 'mitochondria'. From 1964 to 1966 he worked as a postdoctoral fellow with Efraim Racker at the Public Health Research Institute of the City of New York on mitochondrial energy production. After a brief interlude back in Vienna, he emigrated to the USA in 1968 where he accepted a professorship at the Biochemistry Department at Cornell University in Ithaca, NY. Six years later, he moved to what was then the recently established Biozentrum of the University of Basel, which he directed from 1985 until 1987. His research dealt mostly with the mechanism of mitochondrial biogenesis and led to the discovery of mitochondrial DNA and of many key features of mitochondrial biogenesis. He has served as Secretary General of the European Molecular Biology Organization (EMBO) and as member of many scientific advisory bodies around the world. His achievements have been honored with numerous prestigious national and international prizes, honorary doctorates, and admission to scientific academies, including the Royal Swedish Academy of Sciences, the American Academy of Arts and Sciences, and the US National Academy. After his retirement in 2000, the Swiss Federal Government appointed him as president of the Swiss Science and Technology Council. After stepping down from this position in 2004, he became known to a wider public as an essayist and book author. During his years in Austria he also worked as a violinist with the Graz Philharmonic Orchestra, the Graz Opera and the Volksoper in Vienna. He and his Danish wife Merete have three children.

ABOUT THE TRANSLATOR

Andrew Shields was born in Detroit, Mich., in 1964, and was raised in California, Ohio as well as in England. He studied at Stanford University and the University of Pennsylvania, where he received his doctorate in Comparative Literature. He first moved to Europe in 1991 to study in Berlin, and has lived in Basel since 1995, where he teaches at the English Seminar at the University of Basel and also works as a freelance translator. His many translations from German include the correspondence of Martin Heidegger and Hannah Arendt, two volumes of poetry by Dieter M. Gräf, as well as numerous other poems by a wide range of contemporary poets. His own poems have appeared in many journals, as well as in the chapbook *Cabinet d'Amateur* (Cologne, Darling Publications, 2005), and he has turned many of his poems into songs for his band *Human Shields*.

INDEX